Communications
in Computer and Information Science   **2242**

Series Editors

Gang Li , *School of Information Technology, Deakin University, Burwood, VIC,
Australia*
Joaquim Filipe , *Polytechnic Institute of Setúbal, Setúbal, Portugal*
Ashish Ghosh , *Indian Statistical Institute, Kolkata, West Bengal, India*
Zhiwei Xu, *Chinese Academy of Sciences, Beijing, China*

Rationale

The CCIS series is devoted to the publication of proceedings of computer science conferences. Its aim is to efficiently disseminate original research results in informatics in printed and electronic form. While the focus is on publication of peer-reviewed full papers presenting mature work, inclusion of reviewed short papers reporting on work in progress is welcome, too. Besides globally relevant meetings with internationally representative program committees guaranteeing a strict peer-reviewing and paper selection process, conferences run by societies or of high regional or national relevance are also considered for publication.

Topics

The topical scope of CCIS spans the entire spectrum of informatics ranging from foundational topics in the theory of computing to information and communications science and technology and a broad variety of interdisciplinary application fields.

Information for Volume Editors and Authors

Publication in CCIS is free of charge. No royalties are paid, however, we offer registered conference participants temporary free access to the online version of the conference proceedings on SpringerLink (http://link.springer.com) by means of an http referrer from the conference website and/or a number of complimentary printed copies, as specified in the official acceptance email of the event.

CCIS proceedings can be published in time for distribution at conferences or as post-proceedings, and delivered in the form of printed books and/or electronically as USBs and/or e-content licenses for accessing proceedings at SpringerLink. Furthermore, CCIS proceedings are included in the CCIS electronic book series hosted in the SpringerLink digital library at http://link.springer.com/bookseries/7899. Conferences publishing in CCIS are allowed to use Online Conference Service (OCS) for managing the whole proceedings lifecycle (from submission and reviewing to preparing for publication) free of charge.

Publication process

The language of publication is exclusively English. Authors publishing in CCIS have to sign the Springer CCIS copyright transfer form, however, they are free to use their material published in CCIS for substantially changed, more elaborate subsequent publications elsewhere. For the preparation of the camera-ready papers/files, authors have to strictly adhere to the Springer CCIS Authors' Instructions and are strongly encouraged to use the CCIS LaTeX style files or templates.

Abstracting/Indexing

CCIS is abstracted/indexed in DBLP, Google Scholar, EI-Compendex, Mathematical Reviews, SCImago, Scopus. CCIS volumes are also submitted for the inclusion in ISI Proceedings.

How to start

To start the evaluation of your proposal for inclusion in the CCIS series, please send an e-mail to ccis@springer.com.

Gongzhu Hu · Krishna K. Kambhampaty ·
Indranil Roy

Editors

Computer Applications in Industry and Engineering

37th International Conference, CAINE 2024
San Diego, CA, USA, October 21–22, 2024
Proceedings

 Springer

Editors
Gongzhu Hu
Central Michigan University
Mount Pleasant, MI, USA

Krishna K. Kambhampaty
Pennsylvania State University
University Park, PA, USA

Indranil Roy
Southeast Missouri State University
Cape Girardeau, MT, USA

ISSN 1865-0929   ISSN 1865-0937 (electronic)
Communications in Computer and Information Science
ISBN 978-3-031-76272-7   ISBN 978-3-031-76273-4 (eBook)
https://doi.org/10.1007/978-3-031-76273-4

Preface

The 37th International Conference on Computer Applications in Industry and Engineering (CAINE 2024), held October 21–22, 2024, in San Diego, served as a key forum for professionals and academics to discuss new developments in the application of computer technologies across industries and engineering disciplines. This prestigious event, organized by the International Society for Computers and Their Applications (ISCA), featured a rich agenda, including keynote presentations, contributed paper sessions, and a special focus on rewarding outstanding work through the Best Paper Award.

In addition to its general sessions, CAINE 2024 hosted a **Special Session on Advanced Software: Theory and Application**. This session, which highlighted four selected papers, was dedicated to cutting-edge research that bridges theoretical advancements in software engineering with practical, real-world applications. The session fostered dialogue on how theoretical constructs in software development can be applied to enhance system performance, scalability, and security in various industrial contexts. Researchers and practitioners participating in this session explored how innovations in software theory are being integrated into advanced software systems, addressing the challenges of modern computing environments.

In total, **17** full papers were accepted for CAINE 2024 out of a total of 20 submissions. All papers were subjected to a stringent double-blind peer review process with each submission receiving three reviews. Reviewers were drawn from the distinguished CAINE program committee and assigned an average of three papers each. The review process ensured fairness and objectivity. This was made possible by the availability of a diverse pool of qualified reviewers within the committee.

This year's conference pushed the boundaries of software applications, with the special session on advanced software offering valuable insights into both emerging theories and their practical implementations.

September 2024   Krishna K. Kambhampaty
Indranil Roy

Organization

General Chair

Gongzhu Hu   Central Michigan University, USA

Program Committee Chairs

Krishna K. Kambhampaty   Pennsylvania State University, USA
Indranil Roy   Southeast Missouri State University, USA
Takaki Goto   Toyo University, Japan

Program Committee\Reviewers

Abdullah Al-Shoshan   Qassim University, Saudi Arabia
Antoine Bossard   Kanagawa University, Japan
Charitha Hettiarachchi   University of Houston-Clear Lake, USA
Christoph Wunck   Emden/Leer University of Applied Sciences, Germany
Faisal Kabir   Penn State University Harrisburg, USA
Kendall Nygard   North Dakota State University, USA
Masashi Toda   Kumamoto University, Japan
Miroslav Kulich   Czech Technical University in Prague, Czech Republic
Naohiro Ishii   Aichi Institute of Technology, Japan
Norhaslinda Kamaruddin   MARA University of Technology, Malaysia
P. K. Wong   University of Macau, China
Said Ghoniemy   Ain Shams University, Egypt
Reshmi Mitra   Southeast Missouri State University, USA
Ramzi Haraty   Lebanese American University, Lebanon
Sultan Aljahdali   Taif University, Saudi Arabia
Bidyut Gupta   Southern Illinois University-Carbondale, USA
Yan Shi   University of Wisconsin-Platteville, USA
Beckry Abdel-Magid   Winona State University, USA
Bharat Bhushan   Sharda University, India
Sudhir K. Sharma   Institute of Information Technology and Management, India

Contents

Parallel Computing and Algorithms

Data Processing and Image Analysis

Networking and Edge Computing

Cryptography and Pseudorandom Generators

Machine Learning and AI Applications

Development and Optimization of an Ultra-lightweight Deep Spoken Keyword Spotting Model for FPGA Acceleration

Trysten Dembeck[(⊠)] and Chirag Parikh

Grand Valley State University, Grand Rapids, MI 49504, USA
dembeckt@mail.gvsu.edu, parikhc@gvsu.edu

Abstract. Automatic speech recognition (ASR) has become one of the most advanced and studied domains in human-facing machine learning applications. Spoken Keyword Spotting (KWS), a subset of ASR, is a technology that enables systems to detect specific keywords or phrases in spoken language. Modern machine learning models, such as deep neural networks, have significantly advanced the performance and accuracy of KWS systems. However, they often demand substantial computational resources and introduce latencies that limit their real-time applicability and offline use. This has become a tremendous problem where faster and more efficient processing methods dominate and better meet industry demands. To address this challenge, this paper developed a lightweight 1-Dimensional convolutional neural network based on the Mel-frequency cepstral coefficient input and compressed it with quantization and pruning for deployment onto FPGA hardware. The developed model achieved near state-of-the-art performance with far fewer parameters and a simpler architecture than comparable models in literature, and it showed significant model compression with only minor accuracy degradation. This paper also leveraged FPGAs as the hardware deployment strategy to evaluate their effectiveness as inference accelerators for KWS models based on their resource utilization and latency performance improvements.

Keywords: Keyword Spotting · Speech Recognition · Deep Learning · FPGA · Hardware Acceleration · Model Optimization

1 Introduction

Spoken keyword spotting (KWS) is an essential component of contemporary automatic speech recognition (ASR) systems that enables the identification of specific keywords within spoken utterances [1]. It has revolutionized hands-free control over the environment and often manifests as wake-word activation applications in many consumer electronics such as smartphones, home automation devices, wearable technologies, and digital personal assistants [2].

These applications of KWS technologies are possible due to developments in deep learning, model compression, and accelerated computing which have improved the performance of KWS models and has diversified the types of computing systems that the

© The Author(s), under exclusive license to Springer Nature Switzerland AG 2025
G. Hu et al. (Eds.): CAINE 2024, CCIS 2242, pp. 3–20, 2025.
https://doi.org/10.1007/978-3-031-76273-4_1

technology can be deployed to [1]. Yet, the development of KWS systems, and particularly the intricacies of its deployment, is still an evolving and debated field. Therefore, the objectives of this paper are three-fold: to design and assess a 1-D convolutional neural network (CNN) for KWS, to examine the extent that the developed 1-D CNN can be optimized and compressed, and to evaluate the effectiveness of hardware acceleration for KWS by deploying the model on a field-programmable gate array (FPGA).

To accomplish these goals, an efficient, small, and accurate KWS neural network was developed with the intention to deploy it onto an FPGA to mitigate the notoriously long processing times of deep learning models. Each step of the KWS neural network's development—including its training, evaluation, and conversion into an FPGA digital circuit—is described in this paper. The conversion of the 1-D CNN to an FPGA digital circuit was accomplished with a co-design toolset called hls4ml, so a description and analysis of this tool and how it was applied is also described. Additionally, comparisons to other KWS neural network architectures in literature are provided along with an evaluation of inference latencies on various computing devices.

1.1 Spoken Keyword Spotting Task

Spoken keyword spotting involves the classification of specific keywords in streams of audio or recorded utterances [1]. Unlike open-vocabulary ASR systems that attempt to convert any utterance into its equivalent textual format, KWS systems use a finite vocabulary that they identify specific keywords from. KWS systems have a chosen list of words called *target words* or *in-vocabulary* words. All other words (or sounds) are referred to as *out-of-vocabulary (OOV)* and are classified into an additional *Unknown* category. Some KWS models also include a *Silence* category for streams of audio where no sound is present other than background noise. An example of a typical KWS task is shown in Fig. 1.

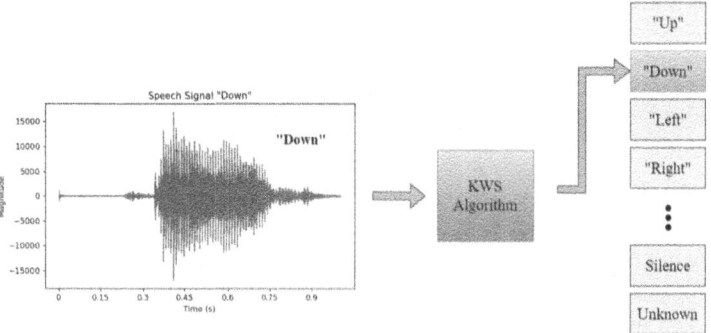

Fig. 1. Generalized Keyword Spotting Task

Behind most modern KWS implementations are deep learning models based on various neural network architectures. Deep learning KWS (deep KWS) models have replaced more classical and rudimentary decision-making algorithms–like Hidden Markov Models (HMMs)–and have greatly improved their abilities to accurately identify keywords

[1]. Various other historical methods have been applied, such as Dynamic Time Warping (DTW) and Large Vocabulary Continuous Speech Recognition (LVCSR) models, but they fail to scale to larger keyword vocabularies and diverse voices or are more complex than necessary for the KWS task [1, 3]. Therefore, the scope of this paper looks past the historical methods and instead focuses on the superior neural network-based approaches that have become profound in literature. This paper specifically implements and evaluates a 1-D CNN to achieve the KWS task and to further the field of deep KWS with lightweight networks.

However, the increased computational complexity of deep learning-based KWS has also imposed strict resource, energy consumption, and latency restrictions for the devices on which they can be implemented. Neural network approaches often require many bytes of storage and hundreds of thousands of multiply-accumulate (MAC) operations which may need enormous amounts of computing cycles to complete [3]. These restrictions have prevented larger and more accurate KWS models from being deployed directly into embedded hardware and forces implementations to look at other methods of edge deployment such as cloud computing [1]. But, advancements in deep learning training methods have mitigated the added complexity problem by developing techniques for compressing large and complex models into a smaller footprint that are compatible with the memory and speed restrictions of embedded systems.

1.2   Accelerating Deep KWS Inference through FPGA Deployment

Despite applying model compression techniques, the processing limitations of devices that use KWS may still incur long inference latencies which diminishes their effectiveness in embedded hardware. Inspired by the acceleration of inference latencies of large machine learning models through parallelization from graphical processing units (GPUs), and more recently tensor processing units (TPUs), this paper leverages FPGAs as the hardware deployment platform [4]. While GPUs and TPUs are far too expensive in both cost and power consumption for embedded systems to utilize, FPGAs may be more suitable as they boast low costs, low power consumptions, high clock speeds, and offer similar or better parallelization opportunities. Based on these characteristics, this paper also aims to determine if FPGAs are effective at overcoming latency constraints in embedded speech recognition systems by providing similar or better performance improvements to GPUs.

Deploying a complex machine learning model onto the resource-limited hardware of an FPGA, and similarly onto general microprocessors, is made possible by applying model compression techniques such as quantization and pruning. Each of these techniques can improve resource utilization and inference latency by synthesizing a more efficient digital circuit on FPGA fabric while minimally affecting the model's accuracy [5]. Both quantization and pruning were performed on this paper's KWS model to allow for a more complex and accurate configuration of it to be deployed. Additionally, an open-source co-design tool for high-level synthesis (HLS), called hls4ml, was used to streamline the conversion of the model and its weights into an FPGA design. This approach shifted the focus of deployment from low-level hardware descriptive language (HDL) and register transfer logic (RTL) levels to an iterative HLS method.

The rest of this paper is structured as follows: The development and training of the deep KWS model, including its architecture, its compression, and its conversion into an FPGA digital circuit is described in section two. In section three, the KWS model is evaluated for its classification accuracy and its inference latency on different hardware platforms including the target FPGA. Additionally, the model's performance is compared to similar state-of-the-art models in literature. Section four highlights the key findings of this research and addresses the broader implications of FPGA-accelerated machine learning tasks.

2 Design Methodology

The design methodology describes how the KWS machine learning model of this study was devised, how it was trained, how it was evaluated, and how it was converted into an FPGA-compatible design. Firstly, the dataset and preprocessing strategies used to train the devised KWS model are discussed. Next, the chosen model architecture is described in relation to its KWS classification task. Finally, the model's conversion into an FPGA-compatible design is explained. It is important to note here that machine learning model development is an iterative process where the specific hyperparameters used at each stage are refined through cross-validation. As such, the specific hyperparameters in each stage of development were determined through cross-validation until the best-performing configuration was found.

2.1 Dataset Selection and Class Balancing

The initial step towards developing a deep KWS model was selecting its training dataset. The model of this paper was trained with the Google Speech Commands V2 dataset [6]. Containing 105,829 one-second-long speech files sampled at 16 kHz and representing 35 words, the Google Speech Commands V2 dataset was designed to assist in the development of speaker-independent limited vocabulary speech recognition systems [6]. From this dataset, 10 keywords were chosen as in-vocabulary while the other 25 were placed in an *Unknown* class. In addition to the spoken utterances, the dataset includes longer audio files of different background noises like a whirring exercise bike and a running faucet among others [6]. A custom program was written to segment these longer noise files into one-second audio clips for training the model to distinguish common background noises as the final *Silence* class. Table 1 summarizes the keywords and categories that the KWS model was designed to classify.

Each of the 10 keywords, the *Unknown* category, and *Silence* category resulted in the model having a 12-class output. The 10 chosen keywords were selected for being commonly used across KWS literature [1]. The 10 keywords were also specifically chosen because they were originally collected to be useful for IoT applications while the other 25 OOV words were chosen because they cover a wide range of phonemes, or perceptually distinct sounds, in the English language [6].

Having as many words with as much phonetic diversity as possible in the OOV selection is generally beneficial to the overall performance of a KWS model. However, in this dataset there is a significant class imbalance from there being far more samples in the

Table 1. Keyword Selection from Google Speech Commands V2

In-Vocabulary (Keywords)	Out-of-Vocabulary (Unknown)			Silence
down	backward	sheila	seven	washing dishes
go	bed	tree	eight	cat noises
left	bird	visual	nine	exercise bike
no	cat	wow		pink noise
off	dog	zero		running faucet
on	follow	one		white noise
right	forward	two		
stop	happy	three		
up	house	four		
yes	learn	five		
	marvin	six		

OOV class than in the keyword classes. It expected that model performance would be negatively impacted by the large imbalance, but also that it could be mitigated by weighting the less-frequent classes or underperforming classes more. Therefore, class weighting techniques were implemented by providing the underrepresented or underperforming classes with more significance based on a combination of the classes' occurrences in the dataset and empirical evidence for helping the underrepresented classes perform better.

Notably, the *Unknown* class contained the most samples by a large margin and was thus weighted the least in relation to the other classes to prevent the model from having a false sense of accuracy. The *Silence* class was also reduced in importance because its feature map signatures are easily distinguishable from any of the keywords, so the model was allowed to focus more on the subtle distinctions between the actual keywords. This paper also took into consideration the bias in words that real KWS systems encounter in their deployment environments. Most words that are heard by the KWS system are typically not any of the keywords. To mitigate this in training, the *Unknown* class was kept larger than all the other keyword classes while ensuring that each dataset partition and training batches had approximately equivalent distributions across each class. Selecting the dataset and distribution of keywords in this manner specified the number of output classes of the model and provided useful information for choosing pre-processing methods.

2.2 Data Pre-processing and Feature Extraction

Selecting the pre-processing and feature extraction methods for the data going into the KWS model was another critical step in its development. These methods need to be capable of extracting the useful latent features in audio signals, be replicable on the target deployment hardware, and be time efficient. To meet these restrictions, Mel-Frequency Cepstral Coefficients (MFCCs) were used as the input features to the model.

MFCCs are commonly used in audio classification tasks for environmental, musical, or other non-speech noises as they effectively represent the spectral information of sounds [7]. They have also been successfully applied to speech classification tasks such as speaker recognition, emotion recognition, and language recognition [8], and are the most widely used feature in modern KWS applications [1]. MFCCs are particularly useful because they warp human speech signals to the Mel-scale which captures spectral information with a non-linear response similarly to how the human ear perceives sound [8]. Additionally, MFCC features retain the temporal context of speech which can improve the ability of KWS systems to classify utterances based on how the signal changes over time [1]. The MFCC features were extracted from speech signals using the pipeline shown in Fig. 2.

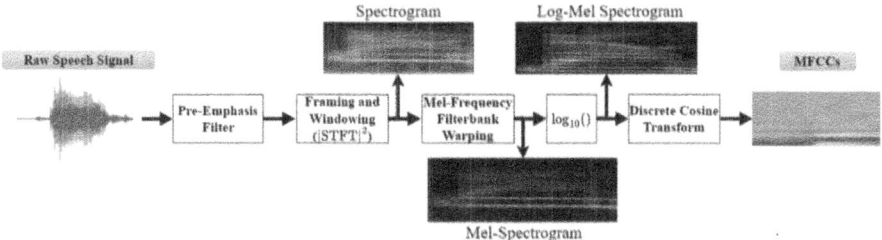

Fig. 2. MFCC Feature Extraction Method

The first stage involves passing the raw speech signal into a pre-emphasis filter. After pre-emphasis, the resulting signal is framed and windowed using a form of the Fast-Fourier Transform (FFT). This research utilized the Discrete Short-Time Fourier Transform (STFT) which has the advantage of preserving the temporal context in the speech signal so that the model can learn from the strong temporal dependencies inherent to human speech. This research also configured the STFT to use 128 bins where each segment comprised of a 16ms window with 50% overlap per step. The Hanning windowing function was also used to reduce spectral leakage between bins, minimize edge effects, and for being computationally efficient. In addition, the magnitude of the STFT was taken and its result was squared to produce a meaningful spectral representation of the speech.

Furthermore, the STFT spectrogram representation was warped with a Mel-Filterbank to convert it into a Mel-Spectrogram. The Mel-Filterbank's parameters were chosen cross-validation to use: 24 Mel bins, a lower edge frequency of 20Hz, and an upper edge frequency of 7,400Hz. The logarithm of the resulting Mel-Spectrogram was taken to produce the Log-Mel spectrogram. Finally, the MFCCs were derived by applying the Discrete Cosine Transform (DCT) on the Log-Mel spectrogram. An example of a speech signal that has been transformed by the defined feature extraction pipeline is shown in Fig. 3.

Propagating a speech signal through the feature extraction pipeline of Fig. 2 produces a 127x24 matrix representing the MFCCs of that signal like the one shown in Fig. 3. The 127x24 shape was used as the input shape of the proposed KWS model of this

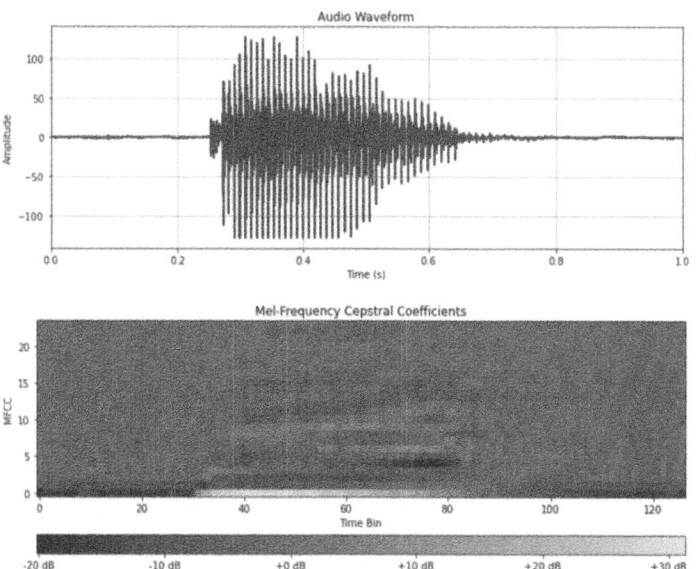

Fig. 3. Speech Signal as its MFCC Features. MFCC features show the alignment of the spectral energy distribution (shown as the lighter and less active regions in the MFCC plot) to regions of voice activity in the utterance.

paper. Having the input feature shape allowed the model architecture to be defined and iteratively improved.

2.3 Model Architecture

The shape of the MFCCs features at the input to the proposed model of this study was critical for determining a model architecture that can effectively learn the relationship between the MFCCs and their associated keyword while still having a lightweight model footprint. Various model architectures have been applied to the KWS task such as fully-connected deep neural networks (DNN), convolutional neural networks (CNN), and recurrent neural networks (RNN). Since MFCCs themselves resemble image-like data, CNNs are a common and effective approach for the KWS task due to their ability to preserve and propagate larger amounts of local context forward through a model [9, 10]. In literature, 1-D and 2D-CNNs, along with their modified versions, are common with each having their own benefits and drawbacks for the KWS task.

   This paper further investigates the use of a 1-D CNN architecture for its computational benefits over the 2-D CNN. Since low-latency inference is prioritized for this paper's model, one benefit of interest is the basic time complexity of propagating an input through a 1-D CNN. For a KxK input shape being convolved by an NxN kernel, 2-D CNNs have a time complexity of $O(N^2K^2)$ while 1-D CNNs have a time complexity of $O(NK)$, making 1-D CNNs less computationally complex and time-complex than 2-D CNNs and therefore providing lower overall baseline inference latencies [10]. In addition, 1-D CNNs can provide reliable performance with very few parameters while typical 2-D CNN applications use much deeper networks and parameter spaces [10]. 1-D CNNs are therefore especially useful for meeting the memory and latency requirements of FPGA deployment.

An additional caveat for FPGA deployment is that the synthesis tool, Vivado HLS, used to synthesize the mathematical operations of the model into a FPGA digital circuit, imposes a restriction that each layer must have less than 4096 parameters to guarantee that it is completely unrolled into available FPGA resources. As such, this study proposes a model comprised of two 1-D CNN layers and two fully connected layers as shown in Fig. 4.

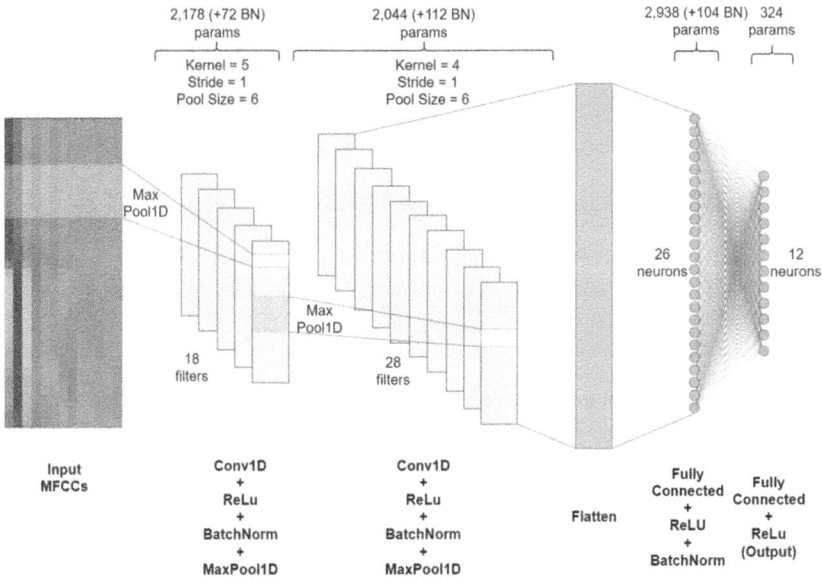

Fig. 4. MFCC-based 1-D CNN KWS Model Architecture

More specifically, the model is first composed of the input layer which takes the 127x24 MFCCs feature map as its input. Following the input stage is a 1-D convolution (Conv1D) layer with rectified linear unit (ReLu) activations. The activations are then followed by a batch normalization layer which was implemented for numerical stability and greatly reduced training times. To complete the Conv1D block, a 1-D maximum pooling (MaxPool1D) layer was included as the dimension reduction strategy.

The first Conv1D layer had 18 filters, a convolving kernel with a width of five, and a stride of one. After the ReLu activation and BatchNorm, the MaxPool1D operation with a pooling size of six was applied to reduce the size of the feature map. The next Conv1D layer followed the same structure as the first but used 28 filters with a kernel size of four and a stride of one. Similarly, after the ReLu activation the BatchNorm operation was applied, a MaxPool1D layer was used to further reduce the dimensions of the propagating feature map.

The following flattening layer was required to make the subsequent fully connected layers compatible with the outputs of the second Conv1D layer. Once flattened, the first fully connected layer used 26 neurons and ReLu activations. Finally, the last fully connected layer has 12 neurons, one for each of the 12 keyword classes, along with a final ReLu activation. The ReLu activation at the output was used in place of a Softmax output to minimize any unnecessary computations being done by the model, further reducing its computational latency. Overall, the model used 7,772 trainable parameters with each layer having less than 4,096 parameters. Having the complete model architecture allows for the training stage to commence with minor adjustments.

2.4 Model Training and Compression

Model Training. The model was trained with the TensorFlow Keras framework in the Python programming language. The dataset was broken into a training split of 80%, a validation split of 10%, and a test split of 10%. The test split was specifically reserved for final evaluation of the model and was never seen by the model during training or validation.

To improve model generalization, additional dropout layers were included in between every hidden layer of the model at a 5% dropout rate to randomly force 5% of all outputs of a layer to zero. This dropout rate is lower than some common training implementations but was strategically and empirically chosen to be lower so that the model can take advantage of the wide range of accents, cadences, pronunciations, pitches, and other speaker variabilities already present in the dataset. Additionally, since the model had so few parameters, higher dropout rates quickly degraded the model's training performance. To further reduce overfitting during the training process, L1 kernel regularization of 0.001 was implemented in every hidden layer model to encourage the model to learn simpler patterns in the data.

In addition, a custom learning rate schedular was applied to half the learning rate every 10 epochs with a starting learning rate of 0.015. The training routine was set to run for 100 epochs using a batch size of 256. The model was trained with the Adam optimizer and used the sparse categorical cross-entropy loss function. Through cross-validation, the previously described model was developed and was further modified to include quantization and pruning.

Model Optimization. Machine learning models that are being deployed into embedded systems can use or require model optimization techniques so that they fit within the device's memory footprint and can run inference with sufficient latency. For this implementation, both quantization and pruning techniques were applied to the proposed KWS model to minimize its memory footprint and latency.

Quantization. During model training, the weights and activations of the model are 32-bit signed floating-point values by default. Reducing the number of bits used to store model weights and perform their mathematical operations, known as quantization, has shown significant reductions in the required memory needed for storage and the energy consumption of models in embedded hardware [11]. Quantization can be performed either during training, called quantization-aware training (QAT), or after training [11]. QAT has the added benefit of being able to regain any lost accuracy through fine-tuning the model and minimizing the loss function at the new bit widths [12]. Generally, selecting the bit widths prior to training is preferred as it also allows for predictive behavior when the model is translated to a hardware accelerator which may be able to internally utilize a specific bit width [12]. The bit widths of each layer and activation can vary within a model and was another hyper-parameter that was carefully tuned through cross-validation.

For simplicity, this paper quantized each hidden layer and its activation to use 12 bits comprised of five integral bits and six fractional bits with 1-bit reserved for the sign. Figure 5 below shows the difference in the baseline parameters and the QAT parameters used for training the model.

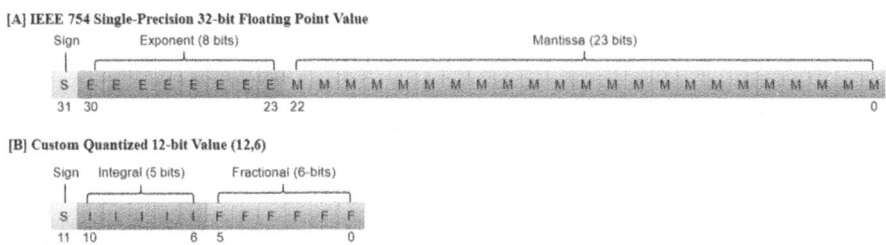

Fig. 5. Baseline Bit Width Compared to Quantized Bit Width. (A) Demonstrates the IEEE 754 floating point representation of parameters used for training the baseline model. (B) shows the 12-bit custom bit width used for quantization-aware training of the optimized model.

The model was trained using QAT which allowed the model to learn the patterns in the training dataset with the lower bit width and improve its accuracy through minimizing the loss function with quantization noise present. To make this compatible with the feature extraction methods, the MFCCs were also extracted using a 12-bit precision. In addition to quantization, pruning is another important optimization technique for deploying models onto embedded hardware.

Pruning. Pruning is a model compression technique that aims to induce sparsity in a model's weights, neurons, filters, or layers by forcing them to zero or removing them from the architecture completely [11]. Sparsity is typically induced in a model by removing model parameters that do not have significant effects on the output or by being significantly lower in magnitude compared to surrounding weights in the layer [13]. Pruning can be done with negligible degradation in model accuracy even when pruning a model to 70–80% sparse making it a critical step in optimizing a model for deployment in an embedded system [13].

This paper pruned the quantized and fine-tuned KWS model's low-magnitude weights to a target sparsity of 50%. An additional 50 epochs were used after QAT to iteratively prune unimportant weights from the model. This was the final step in model optimization as it requires a fully trained model to find weights that do not significantly affect the output. Lastly, the optimized and compressed model was converted into a compatible FPGA digital circuit as described in the next section.

2.5   FPGA Hardware Deployment for Inference Acceleration

The proposed KWS model was converted into an FPGA design using the open source hls4ml package. The hls4ml package provides high-level methods for converting TensorFlow Keras models into ultra-low latency circuits for the target FPGA as shown in their proposed design flow in Fig. 6 [14].

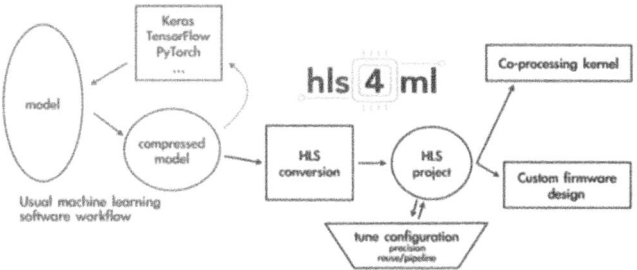

Fig. 6. hls4ml Design Flow (From [14])

The hls4ml framework provides various ways to customize the digital circuit that is generated from the model. These methods directly affect the latency of inference and the usage of important FPGA resources such as BRAM, digital signal processing (DSP) blocks, in addition to the flip-flop (FF) and lookup table (LUT) programmable logic units [14]. Modifying these usage statistics is accomplished with the precision and reuse factors. Choosing the precision of each layer in the model selects the bit width to use for all weights, activations, and multiply-accumulates (MACs) in the FPGA circuit. The reuse factor determines the parallelization of the model in the FPGA fabric by altering how many times a single multiplier (DSP block) is used in performing a multiplication operation. There is a careful balance between throughput and resource usage, particularly when modifying the reuse factor, which needs to be considered when looking at the other requirements of the FPGA design [14].

This study focused on minimizing inference latency and maximizing throughput, so the reuse factor was set to 36 for each layer so that the model was unrolled to the greatest extent while not using every available DSP slice on the FPGA. The precision was set to use the same bit width used during model quantization to ensure the model outputs would exactly match the TensorFlow-trained model's outputs for the same input sequence.

Once these parameters were compiled into the HLS model, hls4ml was invoked to synthesize and implement both KWS models into RTL designs for evaluation and comparison on FGPA hardware. The FPGA hardware used for evaluation in this study was the PYNQ-Z2 system-on-chip (SoC) FPGA. The PYNQ-Z2 houses the Zynq-7020 SoC with a dual-core ARM Cortex-A9 processor that has a well-rounded FPGA fabric directly attached to it. Both models were programmed entirely into the FPGA fabric and interfaced with the processing system to evaluate each model's inference latency described in the next section.

3 Results and Comparisons

This section highlights the results of the developed models of this study by comparing a baseline model and the compressed version of it. It also makes comparisons between other state-of-the-art models in literature. The results of training the baseline model and compressed model are presented to highlight the resulting accuracy and performance of both. In addition, the deployment of both models onto an FPGA were evaluated to quantify the resource usage and inference latency improvements.

3.1 Model Training and Optimization Results

Various types of post-training evaluation methods were applied, such as top-one accuracy, confusion matrices, and a t-distributed stochastic neighbor embedding (t-SNE) metric, to compare the performance of the baseline model and the compressed model. In addition, the models of this paper were compared to the state-of-the-art optimized models in related literature based on similar accuracies and sizes.

Baseline Model vs Compressed Model Performance.

Both the baseline model and the optimized model were trained and evaluated with the same dataset partitions, and both performed well with high accuracies. Table 2 summarizes the results of evaluating both models with the held-out test dataset.

Table 2. Baseline Model and Compressed Model Evaluation

Model	Accuracy (%)	Sparsity (%)	Parameter Count	Size (KB)
Baseline	91.48	0.000	7772	30.36
Optimized	90.16	48.88	7772	11.38

The baseline model achieved a top-one accuracy of 91.48% with 7,772 parameters taking up only 30.36KB in total, and naturally had no sparsity. The compressed and optimized model was fine-tuned to achieve a 90.16% top-one accuracy for a negligible 1.32% accuracy degradation from the baseline model. The target sparsity of the optimized model was 50% but achieved 48.88% sparsity within the given 50 pruning epochs. Adjusted for the 12-bit quantized weights, the optimized model takes up only 11.38KB for a size-reduction factor of 2.67. Moreover, depending on the deployment strategy, zero-valued weights can potentially be completely removed from the model during inference to further reduce the memory map by the factor that the model was pruned to. This additional removal of 48.88% of the model's zero parameters would leave the model at an extremely small size of 5.56KB, although it was not performed during these experiments.

Another method of evaluating the model's performance for each keyword is the confusion matrix, shown in Fig. 7, which displays the distribution of true labels verses the actual predictions for each keyword.

The confusion matrices better show the models' abilities to classify each class correctly. Along the diagonals are the overall top-one accuracies (between 0.0 and 1.0)

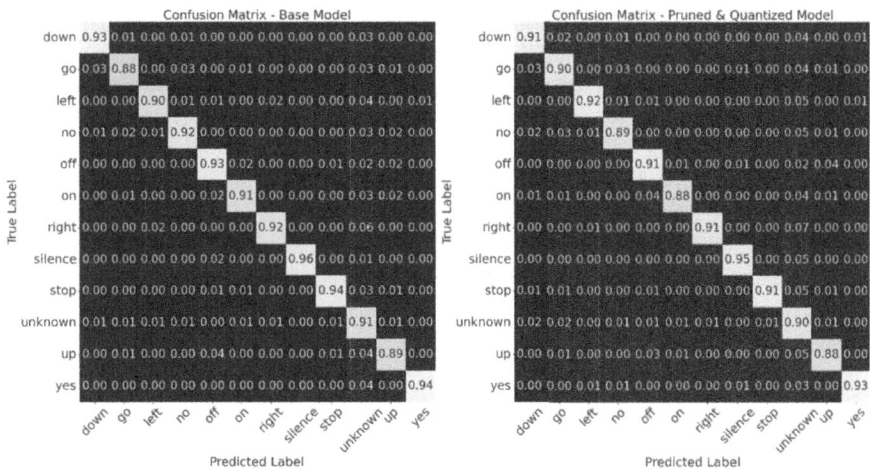

Fig. 7. Confusion Matrices for Base Model and Optimized Model

for each class when evaluated against the test set. Some keywords perform better than others, but both the baseline model and the optimized model have very similar confusion matrices indicated by the minor overall accuracy difference. Other words gained slightly higher classification accuracies in the optimized model with its simpler representation of the feature space. Overall, only minor accuracy degradations were seen between any of the keywords.

In addition to accuracy metrics, a t-distributed stochastic neighbor embedding (t-SNE) metric was applied to the output layer of the optimized model. The resulting t-SNE scatterplots are shown in Fig. 8.

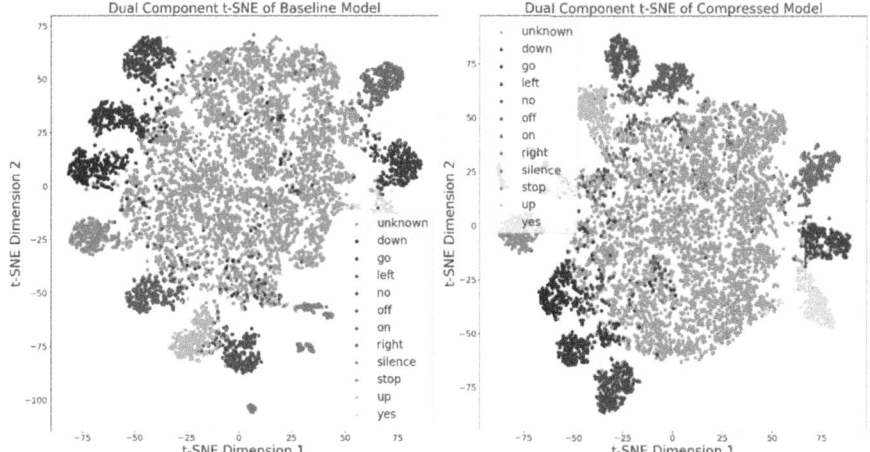

Fig. 8. t-SNE Visualization of Baseline (Left) and Compressed (Right) Model's Output Layers

The t-SNE plots in Fig. 8 demonstrate that both models were able to separate keywords with both high accuracy (indicated by most samples from each class being grouped together) and with good confidence (indicated by the clear separations in groupings between each class). The two configurations also had very similar groupings and indicated that their accuracies were close. Each keyword was placed tightly in its own region with good separation between other keywords indicating that the models could score a particular true keyword with a significantly higher value than the other keywords for most inferences. Notably, the large region of *Unknown* samples was widespread across the plot which indicates the range of phonemes and speaker variabilities present in the dataset. Further comparisons were done to evaluate these KWS models against state-of-the-art models in literature based on their classification accuracies.

Performance of Related Works.

There are many deep KWS models that vary in their neural network architectures, complexities, and performances. Zhang et al. [15] describes many of these different model types and provides metrics for each of them based on the goal of making highly accurate models specifically for embedded system deployment. The models evaluated were comparable in size and performance to the model of this work and are summarized in Table 3 by showing the smallest models trained from each network type.

Table 3. Comparison between this paper's models and Zhang et al.'s [15] comprehensive exploration of different architectures on Google Speech Commands. Model sizes shown are represented as 8-bit weights and activations. The models are ordered by their sizes in Kilobytes.

Model	Accuracy (%)	Size (KB)
This Paper		
1-D CNN 16-bit quantized	90.16	11.38
1-D CNN 32-bit	91.48	30.36
Zhang et al. [15]		
DS-CNN	94.40	38.60
Basic LSTM	92.00	63.30
GRU	93.50	78.80
2-D CNN	91.60	79.00
LSTM	92.90	79.50
CRNN	94.00	79.70
DNN	84.60	80.00

Many of the models from [15] employ recurrent architectures like the Gated Recurrent Unit (GRU), Long Short-Term Memory (LSTM), and the Convolutional Recurrent Neural Network (CRNN). Recurrent networks are better equipped to take advantage of long-term temporal dependencies in the input features in comparison to 1-D CNNs which are restricted to capturing more local temporal dependencies.

However, the baseline 1-D CNN model of this paper was able to competitively achieve an accuracy of 91.48%. Compared to Zhang et al.'s [15] experiments in Table 3, the baseline model exceeded the DNN's accuracy by 6.88% with 62.05% less storage. It also fell just short of the basic LSTM's accuracy by 0.52% but utilized 52.03% less storage. The baseline model also had an accuracy near Zhang et al.'s [15] most accurate model, the depth-separable CNN (DS-CNN), for a difference of 2.92% with this paper's model requiring 21.35% less storage.

The achieved 91.48% performance of this paper's baseline model was accomplished with a meticulously crafted approach and is expected to be much more purpose-built than other practical machine learning models. Its configuration was observed to provide the peak performance that a 1-D CNN of its size could achieve based on a large hyperparameter grid search. Breaking the 92% accuracy mark required models that quickly grew beyond the size of the small footprint models discussed in this comparison, Therefore, the DS-CNN's performance from [15] is highly impressive with such a small size which highlights its superiority at the lightweight KWS task. Overall, these results indicate that the 1-D CNN is capable of being competitive with models that can gather longer-term dependencies but with the 1-D CNN requiring far fewer parameters, especially when optimized with quantization and pruning techniques. After completing the evaluation and comparison of model accuracies, the models were deployed onto FPGA hardware to assess their hardware-accelerated inference latency and FPGA resource utilization.

3.2   FPGA Acceleration Results

The inference latencies of both models were evaluated across three computing devices as summarized in Table 4. This measured the time it took for each model to convert the input features to its predicted output scores and does not include feature extraction time. It is important to note that, within every latency metric, the exhibited performance always depends on the specific characteristics of the deployment platform. Thus, the exact inference latency will vary slightly between different CPUs, GPUs, and FPGAs. This restriction makes it difficult to make direct comparisons of inference latency with other models in literature, so this paper focused on restricted comparisons of the baseline model and compressed model on each hardware architecture.

Table 4. Batched Model Inference Latency Per Sample on Various Device Architectures for 1000 Samples

Device	Baseline Latency (ms)	Compressed Latency (ms)
AMD Ryzen 7 Pro 6850U CPU	4.335	4.335
NVIDIA GeForce RTX 4050 Mobile GPU	2.191	2.191
Pynq-Z2 FPGA	0.374	0.373

The models were specifically evaluated for comparison on a CPU, an NVIDIA GPU, and finally the PYNQ-Z2 FPGA. Prior to evaluating latency, each device was warmed up by classifying a batch of 1000 samples to ensure that all device communication overhead was mitigated and to not skew results. After warmup, the CPU was first evaluated with another 1000 samples which produced an average batched inference latency of 4.335ms per sample for the baseline model. These times were reduced when using an NVIDIA GeForce RTX 4050 Laptop GPU to get an average batched inference latency of 2.191ms per inference. An even greater reduction in time was seen when inference was run on the FPGA which demonstrated a speedup of 11.6 times over the CPU inference and a speedup of 5.9 times over the GPU inference with classifications completing in 373µs, on average. Also, running inference on single samples on the PYNQ-Z2 in real-time with no warmup demonstrated inference latencies between 1.1ms and 1.8ms which still beat the performance of the CPU and GPU even with their batched runs of classifications.

Based on these results, the compression of the model did not have obvious side effects in its inference latency. This can be attributed to two primary reasons. For one, the model only had 7,772 parameters and therefore a comparably low number of MACs to complete, so compressing it further may not have obvious symptoms in inference latency across devices. The second and more dominating reason was that the CPU and GPU tests could not account for the custom 12-bit resolution in the development framework and instead stored the 12-bit values as 32-bit numbers in memory. They also provided no optimization for zero-based multiplications. As for the FPGA implementations, they offered nearly identical latencies primarily due to the same reuse factor being used when synthesizing the design with hls4ml. Despite small differences between the latencies of the baseline and compressed models, these results demonstrate that FPGAs can accelerate the inference of KWS models beyond the ability of GPUs with application-specific hardware implementations.

Finally, FPGA resource utilization was benchmarked for both models to see if optimizing them can lead to improvements. These results are summarized in Table 5.

Table 5. FPGA Resource Utilization of KWS Model on the PYNQ-Z2 FPGA

FPGA Resource	PYNQ-Z2 Total	Base Model Utilization (Count)	Base Model Utilization (%)	Optimized Model Utilization (Count)	Optimized Model Utilization (%)	% Usage Reduction w/ QAT
Slice LUTs	53200	25762	48.42	25496	47.92	1.03
Slice Registers	106400	41867	39.35	38996	36.65	6.86
Slices	13300	11309	85.03	10831	81.44	4.23
BRAM Tiles	140	53	37.86	49	35.00	7.55
DSPs	220	93	42.27	92	41.82	1.08

From Table 5, it was observed that the compressed model did, as expected, provide a reduction in resource usage across each category over the baseline model. Overall, these utilization reductions were lower than expected based on the degree of compression in the optimized model. The small reductions are largely due to the model having so few parameters which likely left little room for FPGA synthesis optimizations. However, utilization could be further improved by better tuning the reuse and quantization parameters in the hls4ml configuration, or by using additional synthesis and implementation-based optimizations in Vivado HLS.

Notably, although both models were implemented with the same reuse factor, the compressed model utilized one less DSP than the baseline model which was likely due to the zero-based multiplications being implemented with simpler digital circuitry and therefore saving one DSP unit. Ultimately, FPGAs proved successful in greatly accelerating the KWS model developed in this paper, and the compression techniques provided slightly better resource utilization.

4   Conclusion

This paper developed and described a lightweight KWS model that achieved a baseline accuracy of 91.48% with only 7,772 weights, or 30.36KB of storage with 32-bit parameters. The model, trained using Google Speech Commands V2 dataset, used a 1-D CNN architecture to extract short-term temporal dependencies in MFCC input features. A compressed version of the model was developed using quantization and pruning techniques. This reduced the model's parameter bit width to 12-bits and resulted in model that was 48.88% sparse. The large reduction in model size exhibited a minor accuracy degradation of only 1.32%. These performance and size metrics compared well with other lightweight models in reviewed literature and had up to a 62.05% reduction in model size when compared to some of the smallest and best-performing models.

In addition, FPGAs were utilized as the model deployment platform to evaluate their effectiveness in accelerating the inference of KWS models. Converting the model into an FPGA circuit and deploying it resulted in an average batched inference latency of 373μs per sample which corresponds to large speedups of up to 11.6 times and 5.86 times over the CPU and GPU latencies, respectively. This research demonstrated that accurate and fast KWS models can be achieved with very few parameters using class balancing techniques along with cross-validating feature extraction hyperparameters and employing model optimization techniques. The successful deployment of the lightweight KWS model onto FPGA hardware is also an indicator of the potential for drag-and-drop KWS integrated circuits that can add voice-control capabilities to a larger variety of embedded systems and applications.

Disclosure of Interests. The authors have no competing interests to declare that are relevant to the content of this article.

References

1. López-Espejo, I., Tan, Z.H., Hansen, J., Jensen, J.: Deep spoken keyword spotting: an overview. IEEE Access **10**, 4169–4199 (2022). https://doi.org/10.1109/ACCESS.2021.313 9508

2. Hoy, M.: Alexa, Siri, Cortana, and more: an introduction to voice assistants. Med. Ref. Serv. Q. **37**(1), 81–88 (2018). https://doi.org/10.1080/02763869.2018.1404391
3. Li, S., Li, J., Han, J., Zhi, T.: Overview of speech keyword recognition technology. J. Phys. Conf. Ser. **1827**, 012013 (2021). https://doi.org/10.1088/1742-6596/1827/1/012013
4. Nikolić, G. S., Dimitrijević, B. R., Nikolić, T. R., Stojcev, M. K.: A survey of three types of processing units: CPU, GPU and TPU. In: 57th International Scientific Conference on Information, Communication and Energy Systems and Technologies (ICEST), pp. 1–6. Ohrid, North Macedonia (2022). https://doi.org/10.1109/ICEST55168.2022.9828625
5. Wang, C., Luo, Z.: A review of the optimal design of neural networks based on FPGA. Appl. Sci. **12**(21), 10771 (2022). https://doi.org/10.3390/app122110771
6. Warden, P.: Speech commands: a dataset for limited-vocabulary speech recognition. Google Brain (2018). https://doi.org/10.48550/arXiv.1804.03209
7. Gourisaria, M.K., Agrawal, R., Sahni, M., Singh, P.K.: Comparative analysis of audio classification with MFCC and STFT features using machine learning techniques. Discov. Internet Things **4**, 1 (2024). https://doi.org/10.1007/s43926-023-00049-y
8. Abdul, Z.K., Al-Talabani, A.K.: Mel frequency cepstral coefficient and its applications: a review. IEEE Access **10**, 122136–122158 (2022). https://doi.org/10.1109/ACCESS.2022.3223444
9. Abdel-Hamid, O., Mohamed, A.R., Jiang, H., Deng, L., Penn, G., Yu, D.: Convolutional neural networks for speech recognition. IEEE/ACM Trans. Audio, Speech, Lang. Proce. **22**(10), 1533–1545 (2014). https://doi.org/10.1109/TASLP.2014.2339736
10. Kiranyaz, S., Avci, O., Abdeljaber, O., Ince, T., Gabbouj, M., Inman, D.J.: 1D convolutional neural networks and applications: a survey. Mech. Syst. Signal Process. **151**, 107398 (2021). https://doi.org/10.1016/j.ymssp.2020.107398
11. Choudhary, T., Mishra, V., Goswami, A., Sarangapani, J.: A comprehensive survey on model compression and acceleration. Artif. Intell. Rev. **53**, 5113–5155 (2020). https://doi.org/10.1007/s10462-020-09816-7
12. Guo, Y.: A Survey on Methods and Theories of Quantized Neural Networks (2018). https://doi.org/10.48550/arXiv.1808.04752
13. Yeom, S.K., Seegerer, P., Lapuschkin, S., Binder, A., Wiedemann, S., Müller, K.R., et al.: Pruning by explaining: a novel criterion for deep neural network pruning. Pattern Recogn. **115**, 107899 (2021). https://doi.org/10.1016/j.patcog.2021.107899
14. Duarte, J., Han, S., Harris, P., Jindariani, S., Kreinar, E., Kreis, B., et al.: Fast inference of deep neural networks in FPGAs for particle physics. J. Instrum. **13**, 7027 (2018). https://doi.org/10.1088/1748-0221/13/07/P07027
15. Zhang, Y., Suda, N., Lai, L., Chandra, V.: Hello Edge: Keyword Spotting on Microcontrollers. ArXiv (2017). https://doi.org/10.48550/arXiv.1711.07128

Hybrid AI Techniques for Non-invasive Fault Detection with Experimental Validation

Hoon Lee[✉] and Ka.C Cheok

Electrical and Computer Engineering Department, Oakland University, Rochester, MI 48309, USA
{hlee23456,cheok}@oakland.edu

Abstract. This paper will provide a non-invasive fault detection solution with Artificial Intelligence (AI) techniques. Non-invasive methods would not require system and machine modification. This solution would use the existing system to collect visual, audio, and vibration data for diagnostics. Because the early signs of fault are numerous, subtle, complex, and difficult to classify and detect, with mathematics and signal processing, the collected diagnostic data will be processed and analyzed for unique patterns and features to be used as input to AI tools. These features will be used to train the AI tools. The subtle characteristics are learned through training; when completed, the trained AI tools can detect them in real-time. Each collected data will be able to detect different faults than others. Some early stages of faults are better detected and diagnosed by visual images than by vibration or audio. Others are better with audio because the cause of the fault is embedded deep inside a machine. For example, an Instrumental Panel (IP) showing engine rpm, and a speedometer complemented by engine sound and wheel vibrations can reveal non-obvious anomalies that might not be shown on the IP. Sound and vibration would be able to provide the early telltale signs of anomalies inside the engine and wheels. These theories are experimented with and validated on an actual vehicle and will show that a non-invasive fault detection solution is a viable solution to early fault detection.

Keywords: Non-invasive · fault-detection · hybrid AI · fuzzy inference system

1 Introduction

Machines are complex equipment developed with many integrated parts and technology, and as technology has advanced as described in journal articles [1], so have the complexity and features. More components and functionalities are installed as more features and requirements are added to the machines. This has increased the complexity and thus the difficulty of maintaining them.

The journal articles [2] investigated how one form of sign could give us methods to recognize faults but there is a lack of investigation on techniques for combining and blending the different telltale signs to improve the diagnostic results. A hybrid of different diagnostic algorithms could extract the essential fault features and blend them to get the best possible diagnostic.

G. Hu et al. (Eds.): CAINE 2024, CCIS 2242, pp. 21–29, 2025.
https://doi.org/10.1007/978-3-031-76273-4_2

There are many fault detection methods for using various types of sensors. In this paper, we will demonstrate how hybrid AI techniques combine these three sensors and methods to perform non-invasive diagnostics methods.

2 Fault Detection and Maintenance Methods

2.1 Types of Fault Detection and Maintenance Methods

There are three known types of fault detection and management methods used. The first and simplest method is Reactive Maintenance (RM). This reactive fault detection and maintenance method would wait until the machine breaks down and is replaced.

The second and most common method is Preventive Maintenance (PM). This time interval method is used to replace a machine promptly before it breaks down.

The last method is Predictive Maintenance (PreM). This is the most desired maintenance method that could predict when the machine is likely to fault with high probability. This method would require monitoring of the signals and detecting subtle and inconspicuous tell-tale signs of faults from the machine [3].

2.2 Difficulty with PreM Implementation

Closed-loop System Modification. Modifying closed-loop systems and machines that are sealed and operational could be difficult. For example, nuclear reactor systems require difficult activities to modify a machine to add new PreM features. We should consider implementing this in a non-invasive approach that would not require any kind of modification to the operating system [4].

Difficult to Detect Faults. It is difficult to detect early signs of faults because faults are exposed from the combination of various factors. For example, visually a machine could look fine, but it could be making unknown noises. Or it might not be making a noise, but you could feel emulating heat from the machine. The signs are conspicuous and subtle and have different variations, features, and characteristics that are difficult to recognize. This system would require collecting and processing data from combinations of sensors to better detect early faults.

Large Data Processing. To be able to continuously record and process large amounts of data and recognize subtle fault features is difficult with only math and signal processing methods. Such a large amount of data would require smart batch processing tools and methods implemented with signal processing techniques. This would require AI tools [5, 6].

3 Proposed Tools

3.1 Non-invasive Fault Detection with AI (NIFDAI)

Non-invasive. This proposed solution is non-invasive, meaning no modification to the operating system is necessary [4]. This external system with trained AI tools would record raw data in the surroundings and detect faults.

Multiple Sensors. This system would have visual, and audio sensors and devices that will record a combination of ambient data. A combination of camera, microphone, and IMU is embedded to monitor the ambient machine surroundings.

AI Tools. This proposed solution is embedded with a hybrid of GoogleNet and LSTM for visual fault detection, Yam-net for audio fault detection, and Long-Short-Term-Memory (LSTM) for 3-axis vibration fault detection. A Fuzzy Inference System (FIS) is used to determine the fuzzy predicting values from different devices and output the most likelihood of the fault.

4   Setup and Simulation of Proposed Solution

4.1   Selected Faults for Simulation

Battery, Fuel Injection, and Starter faults are recorded, simulated, and validated using a sedan vehicle. The faults are recorded with image data, sound data, and 3-axis acceleration data.

4.2   Detection of Faults with Visual Data

Visual data from faults are recorded with a camera and used to train the hybrid of GoogleNet with LSTM and output VisualFaultNet (VisFN). Trained VisFN can recognize the features and image characteristics shown on the recorded vehicle instrumental panel. All recorded fault data are recorded and processed as in the figure below.

Fig. 1. Visual Fault Detection with Camera

4.3   Detection of Faults with Audio Data

Audio data from faults are recorded with a microphone and used to train Yam-net and output AudioFaultNet (AudFN). A trained AudFN can recognize the features and sound characteristics recorded from the vehicle cabin. All recorded fault data are processed as shown in the figure below.

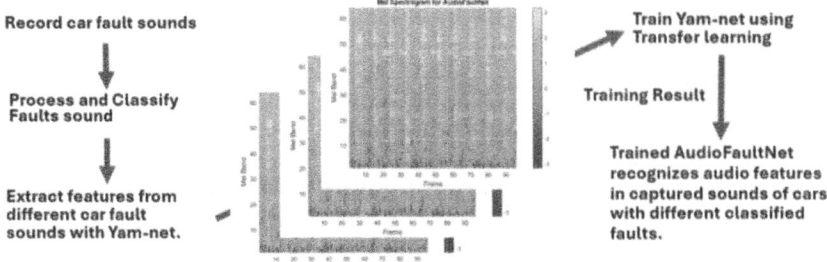

Fig. 2. Audio Fault Detection with Microphone

4.4   Detection of Faults with Vibration Data

3-axis data from faults are recorded with an IMU and used to train LSTM to output Vib-FaultNet (VibFN). Trained VibFN can recognize the features and 3-axis values recorded from the vehicle. All faults' data are recorded and processed as shown in the figure below.

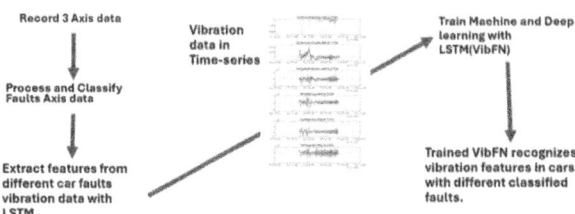

Fig. 3. 3 Axis Fault Detection with IMU

5   Simulation Results

The simulation results are from recognizing and differentiating the faults' features through training, as shown in Fig. 1, Fig. 2, and Fig. 3. The mini-batch prediction techniques with VisFN resulted in a confusion matrix accuracy of 85%. AudFN recognized and differentiated the audio features in faults in a confusion matrix with an accuracy of 78%. VibFN with time series axis data, resulting in confusion matrix accuracy of 82%.

6   Hybrid Decision-Making (DM)

6.1   Fault DM with Fuzzy Logic

Classification. The VisFN is a hybrid-net classifier that uses a trained convolutional neural network (CNN) to process the video frames from the camera looking at the instrument panel and recognize and classify a fault that may show up in the gauges. If fault A shows up, it is labeled as A_c; if B shows up then B_c; and C_c, and so on.

The AudFN is a YamNet classifier that uses CNN to process the audio signals from the microphone listening to the sound from the car and recognize and classify a faulty noise that may show up. If fault A shows up, it is labeled as A_m; if B shows up then B_m; and C_m, and so on.

The VibFN is an LSTM classifier that uses a Recurrent Neural Network (RNN) to process the vibration signals from the IMU monitoring to the vibration from the car and recognize and classify a vibration fault that may show up. If fault A shows up, it is labeled as A_i; if B shows up then B_i; and C_i, and so on (Table 1).

This study considers the outputs A_c, B_c, and C_c to be confidence values taking on real numbers between zero and one. That is A_c, B_c, C_c,... $\in [0,1] \subset R$. Similarly, A_m, B_m, $C_m \in$,... $[0,1] \subset R$ and A_i, B_i, C_i,... $\in [0,1] \subset R$.

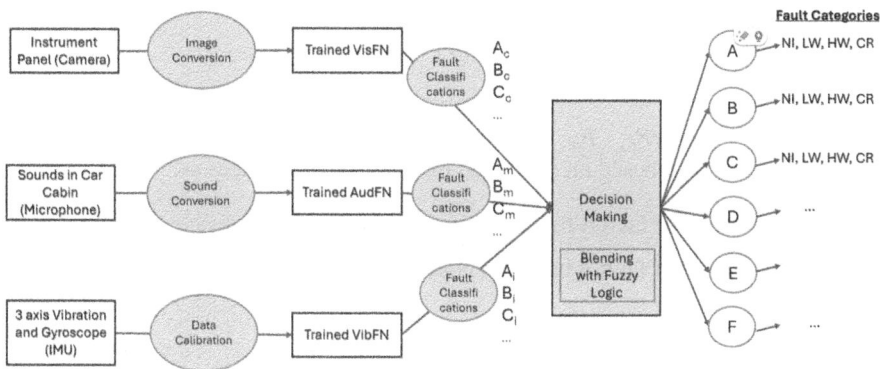

Fig. 4. Decision-making with FIS

Figure 4 shows the outputs of the fault classifiers, and the table below shows the classifications.

Table 1. Classifications of Faults and Devices

Devices	"A" Faulty Battery	"B" Faulty Fuel Injection	"C" Faulty Starter
Camera	A_c	B_c	C_c
Microphone	A_m	B_m	C_m
IMU	A_i	B_i	C_i

The types of faults are labeled as A, B & C. The detection devices as c, m & i. The association variables reported by the devices as:

$$A_c \triangleq \text{variable for fault } A \text{ reported by device } c$$
$$A_m \triangleq \text{variable for fault } A \text{ reported by device } m$$
$$A_i \triangleq \text{variable for fault } A \text{ reported by device } i$$

$$\vdots$$

$$B_c \triangleq \text{variable for fault } B \text{ reported by device } c$$
$$B_m \triangleq \text{variable for fault } B \text{ reported by device } m$$

$$\vdots$$

$$C_c \triangleq \text{variable for fault } C \text{ reported by device } c$$

where
$$A_c, \quad A_m, \quad A_i, \quad \cdots \quad, \quad B_c, \quad B_m, \quad B_i, \quad \cdots \quad, \quad C_c, \quad C_m, \quad C_i, \quad \cdots \quad \in [0, 1] \quad \subset \mathbb{R}.$$
The likelihood of fault based on combined reports from all devices as:

$$L_A \triangleq \text{likelihood of fault } A \text{ based on combined report from device } c, \ m, \ i, \ \cdots$$

$$L_B \triangleq \text{likelihood of fault } B \text{ based on combined report from device } c, \ m, \ i, \ \cdots$$

$$L_C \triangleq \ \cdots$$

where L_A, L_B & $L_C \in [0, 1]$ $\subset \mathbb{R}$

The fuzzy values and degree of association or membership function for a device can be illustrated as follows. Suppose that x is the value of the fuzzy variables of a particular device and the association values are $\mu_{dev,low}(x)$, $\mu_{dev,med}(x)$ & $\mu_{dev,high}(x)$. For example,

$$\mu_{dev,low}(x) = \frac{1}{1 + e^{-m_1(x-c_1)}} \qquad m_1 = -15 \qquad c_1 = 0.3$$
$$\mu_{dev,med}(x) = e^{-m_2(x-c_2)^2} \qquad m_2 = 15 \qquad c_2 = 0.5 \qquad (1)$$
$$\mu_{dev,high}(x) = \frac{1}{1 + e^{-m_3(x-c_3)}} \qquad m_3 = 15 \qquad c_3 = 0.7$$

The fuzzy rules for device fault level are.

Rule 1 If A_c is $\mu_{A_c,low}$ and A_m is $\mu_{A_m,low}$ and A_c is $\mu_{A_i,low}$, then $L_A = w_1$

Rule 2 If A_c is $\mu_{A_c,low}$ and A_m is $\mu_{A_m,low}$ and A_c is $\mu_{A_i,med}$, then $L_A = w_2$

Rule 3 If A_c is $\mu_{A_c,low}$ and A_m is $\mu_{A_m,low}$ and A_c is, $\mu_{A_i,high}$ then $L_A = w_3$

Rule 4 If A_c is $\mu_{A_c,low}$ and A_m is $\mu_{A_m,med}$ and A_c is $\mu_{A_i,low}$, then $L_A = w_4$

Rule 5 If A_c is $\mu_{A_c,low}$ and A_m is $\mu_{A_m,med}$ and A_c is $\mu_{A_i,med}$, then $L_A = w_5$

Rule 6 If A_c is $\mu_{A_c,low}$ and A_m is $\mu_{A_m,med}$ and A_c is, $\mu_{A_i,high}$ then $L_A = w_6$

$$\vdots$$

Rule 27If A_c is $\mu_{A_c,med}$ and A_m is $\mu_{A_m,high}$ and A_c is, $\mu_{A_i,high}$ then $L_A = w_{18}$ $L_A = w_{18}$

The same Rule 1 to Rule 27 applies to Fault B and Fault C. The strength of each rule using multiplication arithmetic for the AND operator

$$
\begin{aligned}
s_1 &= \mu_{A_c,low} * \mu_{A_m,low} * \mu_{A_i,low} \\
s_2 &= \mu_{A_c,low} * \mu_{A_m,low} * \mu_{A_i,med} \\
s_3 &= \mu_{A_c,low} * \mu_{A_m,low} * \mu_{A_i,high} \\
s_4 &= \mu_{A_c,low} * \mu_{A_m,med} * \mu_{A_i,low} \\
s_5 &= \mu_{A_c,low} * \mu_{A_m,med} * \mu_{A_i,med} \\
s_6 &= \mu_{A_c,low} * \mu_{A_m,med} * \mu_{A_i,high} \\
&\vdots \\
s_{27} &=
\end{aligned} \tag{2}
$$

The Sugeno-style Inference output using the weighted average is given by

$$
L_A = \frac{\sum_{k=1}^{27} (s_k * w_k)}{\sum_{k=1}^{27} s_k} \tag{3}
$$

Similar computations will yield the combined fault level L_A. The weighted average as described in the Eq. (3), the L_A, L_B & L_C are the resulting probability of fault from A to C. The higher the value, the more likelihood that is the true fault than other lower value faults. For example, if L_A has the highest value then this would indicate that this is most likely a fault caused by a bad battery.

7   Experiments

7.1   Devices Setup and Descriptions

For illustrations shown in Fig. 5 below, three devices (Camera, Microphone, IMU) are installed to capture visual, sound, and vibration data. The physical location of installed devices is shown as follows:

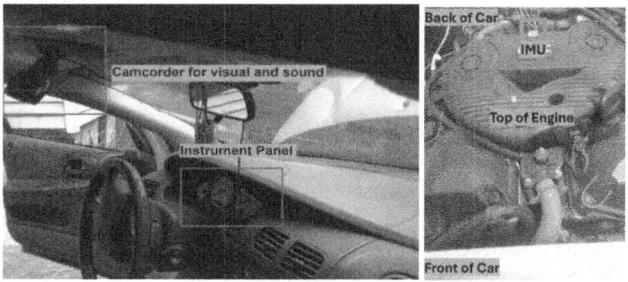

Fig. 5. Devices locations

7.2 Experiments Cases

The experiment cases of A, B, and C produce almost identical visual, sound, and vibrations that are difficult to detect. This experiment is to show the results of NIFDAI detecting the features in image sequences, sound spectrum, and vibration data sequences and predict the fault using FIS. Below are descriptions of starting fault scenarios.

The car not starting due to "A" faulty battery. The vehicle is installed with a damaged and depleted battery. It is charged using cables to the car battery to 13V before experiments.

The car not starting due to "B" faulty fuel injection. The vehicle relay for the fuel pump is removed. This will prevent fuel from being supplied to the engine. The car will start on the first attempt and will immediately halt.

The car not starting due to "C" faulty starter. The vehicle relay for the starter is removed. This will prevent the starter from engaging with gear.

8 Experimental Validation

8.1 Fault Cases

The number of predictions is shown in Table 2 below, on the fault classifications of faults. The fault "A" had the highest recognition of 87 using the camera. Microphone recordings processed through Yam-net showed a value of 79. The fault "C" with IMU had the highest value of 66 as a Faulty Starter. Because the likelihood values of fault "A" is higher than "B" and "C". The resulting value of L_A. is highest and predicted L_A to be the predicted fault.

Table 2. Validation Results on Faulty Battery

Devices	"A" Faulty Battery	"B" Faulty Fuel Injection	"C" Faulty Starter
Camera	87	8	5
Microphone	79	15	6
IMU	3	31	66

9 Conclusions

The validation of the NIFDAI system proposed by this paper shows recognition of fault classification of "A" Faulty Battery. The highest number of 87 probability of correct classification using the camera and other devices validated that this system could detect faults in its recorded features.

The challenging aspect of the NIFDAI system is locating the best location for recording sensors. As this is a non-invasive system, no part is opened or modified, and the sensors are installed while the vehicle is operational. Depending on the location of the device, there were distortions and undesired noise to the recorded data. This has led to outputting incorrect results and required filtering and calibrations.

Future actions would include more devices, and AI tools collecting more data on other fault types and training the AI network models to detect and recognize the faults in real-world cases. There is a lack of shared recorded data on machine faults and collecting and sharing this would advance the diagnostic field of NIFDAI.

References

1. Wolff, J.: How is technology changing the world, and how should the world change technology? section: technology and global change. Glob. Perspect. **2**(1), 27353 (2021). https://doi.org/10.1525/gp.2021.27353
2. Reñonesa, A., et al.: F.A.I.R. open dataset of brushed DC motor faults for testing of AI algorithms. ADCAIJ: Adv. Distributed Comput. Artif. Intell. **9**(4), 83–94 (2020). eISSN: 2255–2863. https://doi.org/10.14201/ADCAIJ2020948394
3. Paul, R., et al.: Conceptualisation of a novel technique to incorporate artificial intelligence in preventive and predictive maintenance in tandem. Mater. Today Proc. **66**, Part 9, 3814–3821 (2022). ISSN 2214–7853, https://doi.org/10.1016/j.matpr.2022.06.250
4. Alotaibi, M., et al.: Non-invasive inspections: a review on methods and tools. Sensors (Basel). **21**(24), 8474 (2021). https://www.ncbi.nlm.nih.gov/pmc/articles/PMC8705398/
5. Dalzochio, J., et al.: Machine learning and reasoning for predictive maintenance in Industry 4.0: Current status and challenges. Comput. Ind. **123**, 103298 (2020). ISSN 0166-3615. https://doi.org/10.1016/j.compind.2020.103298
6. Thyago, P., et al.: A systematic literature review of machine learning methods applied to predictive maintenance. Comput. Ind. Eng. J. **137,** 106024 (2019). https://www.elsevier.com/locate/caie

An Interactive Question Answer Based System on Alzheimer's Disease Using Retrieval Augmented Generation

Sujoy Sen[1], Samay Sarkar[1], Partha Ghosh[2] , Takaaki Goto[3], and Soumya Sen[1(✉)]

[1] University of Calcutta, Kolkata, India
iamsoumyasen@gmail.com
[2] Academy of Technology, Adisaptagram, West Bengal, India
[3] Toyo University, Saitama, Japan
tg@gotolab.net

Abstract. Alzheimer's Disease (AD) presents profound challenges to healthcare systems worldwide, necessitating efficient access to accurate information for optimal care delivery. Effective management of AD requires timely access to accurate information spanning disease etiology, diagnosis, treatment options, and caregiving strategies. However, the vast and constantly evolving body of AD-related literature poses a considerable barrier to efficient information retrieval, particularly for healthcare professionals operating in time-constrained environments. This paper outlines the objectives of a specialized Retrieval-Augmented Generation (RAG) system designed for answering questions related to Alzheimer's Disease (AD), employing prompt engineering and utilizing Pinecone as the vector database. With the goal of enhancing accessibility and comprehension of AD-related information, the system aims to efficiently retrieve relevant data from diverse sources and generate contextually relevant answers tailored to user queries. By leveraging advanced techniques in prompt engineering and vector similarity search, the RAG system empowers healthcare professionals, patients, and caregivers with timely access to accurate and comprehensive information, ultimately facilitating informed decision-making and improving patient outcomes in AD management.

Keywords: Alzheimer's Disease · Question Answering · Retrieval-Augmented Generation · Prompt Engineering · Pinecone · Vector Database

1 Introduction

Alzheimer's Disease (AD) presents an escalating challenge in modern healthcare, characterized by its progressive neurodegenerative effects, cognitive decline, and profound societal impact. With an aging global population, the prevalence of AD continues to surge, projecting a doubling of cases by 2050, thereby amplifying the strain on healthcare systems worldwide.

Alzheimer's disease is a progressive neurodegenerative disorder that predominantly impacts memory, cognition, and behaviour. It stands as the leading cause of dementia

G. Hu et al. (Eds.): CAINE 2024, CCIS 2242, pp. 30–40, 2025.
https://doi.org/10.1007/978-3-031-76273-4_3

in older adults, responsible for 60–80% of cases. The disease is marked by the build-up of amyloid plaques and tau tangles in the brain, which results in the death of nerve cells and subsequent cognitive decline. Initial symptoms often include minor memory lapses, especially concerning short-term memory, and gradually escalate to significant difficulties in language, reasoning, and spatial navigation. As the disease advances, individuals may find it challenging to perform basic daily activities and eventually may require full-time care.

The exact cause of Alzheimer's disease is not yet fully understood, but several risk factors have been identified, including age, genetics, and lifestyle choices. Age is the most significant risk factor, with the likelihood of developing Alzheimer's increasing substantially in individuals over the age of 65. While there is no cure for Alzheimer's, treatments are available to manage symptoms and improve quality of life. These include medications that help regulate neurotransmitter activity and behavioural therapy to address mood and behaviour changes. Researchers continue to investigate potential treatments and preventative measures, aiming to better understand the disease and develop more effective interventions.

Amidst this burgeoning crisis, the application of cutting-edge methodologies becomes imperative. Retrieval-Augmented Generation (RAG), a paradigm that integrates retrieval-based techniques with generative models, emerges as a promising avenue for revolutionizing Alzheimer's Disease management. By harnessing the power of large-scale data repositories and advanced natural language processing (NLP) models, RAG offers a novel framework for synthesizing information, facilitating decision-making, and optimizing patient care pathways in the realm of AD.

At its essence, the RAG framework for Alzheimer's Disease management revolves around the synergistic interplay of two fundamental components:

Retrieval: The retrieval aspect of RAG involves the systematic aggregation and extraction of pertinent data from diverse sources, including electronic health records (EHRs), medical literature, imaging databases, and patient registries. Leveraging sophisticated information retrieval (IR) techniques, healthcare practitioners can access a wealth of structured and unstructured data, encompassing clinical assessments, biomarker profiles, genetic predispositions, and treatment histories. This comprehensive data retrieval lays the groundwork for informed decision-making, personalized risk stratification, and targeted intervention planning in Alzheimer's Disease management.

Generation: Complementing the retrieval phase, the generation component of RAG encompasses the utilization of generative models, such as language models and deep learning architectures, to synthesize contextually relevant insights, recommendations, and prognostications. Through advanced natural language generation (NLG) techniques, RAG facilitates the creation of tailored care plans, patient education materials, and caregiver support resources, thereby empowering stakeholders with actionable information and enhancing communication within the healthcare ecosystem.

The integration of retrieval-augmented generation in Alzheimer's Disease management is inherently data-centric, leveraging large-scale datasets, ontologies, and knowledge graphs to fuel model training and inference. By capitalizing on state-of-the-art AI technologies, including transformers, attention mechanisms, and reinforcement learning algorithms, RAG enables the seamless integration of clinical expertise, empirical evidence, and patient preferences into a unified decision-support framework.

In this documentation, we embark on a comprehensive exploration of the RAG paradigm in the context of Alzheimer's Disease management. Drawing upon interdisciplinary insights from computational linguistics, biomedical informatics, and clinical neuroscience, we elucidate the theoretical foundations, methodological intricacies, and practical applications of RAG in optimizing patient outcomes, enhancing clinical workflows, and driving innovation in AD care delivery.

As we navigate the evolving landscape of Alzheimer's Disease management in an era of data-driven healthcare transformation, the adoption of RAG represents a paradigm shift towards precision medicine, personalized care delivery, and collaborative decision-making. By embracing the synergistic potential of retrieval-augmented generation, we endeavour to mitigate the burden of Alzheimer's Disease, empower individuals and families affected by the condition, and pave the way towards a brighter future for dementia care.

The organization of the paper is outlined as follows: Sect. 2 covers the related studies, followed by the research objectives in Sect. 3. Sect. 4 explains the relevant terminologies. The proposed methodology is detailed in Sect. 5. In Sect. 6, we conduct a case study to evaluate the efficiency of the proposed model. Lastly, Sect. 7 presents the conclusion.

2 Related Work

Alzheimer's disease is one of the threats to the human life as more and more people are being affected over the time. Modern lifestyle, food habits, psychological issues, genetic issues are believed to be the different reasons behind this disease. The exact reasons or trigger points are not known and there is no treatment that can cure Alzheimer's totally. Treatments are done to manage symptoms and to slowdown the advancement of the disease. Use of modern day technology is helping the treatment process of Alzheimer's. Early detection of the disease is one of the key issues of the treatment process.

[1] presents a comprehensive survey on different deep learning based methods for Alzheimer's disease detection. A simple neural network is used to detect the Alzheimer's disease analysing cerebral MRI [2]. Classification of Alzheimer's disease [3] and its prodormal stage, MCI (Mild Cognitive Impairment) helps to prevent progress of memory impairment hence improving the life quality of Alzheimer's patients. These research articles are helpful for those actively involved in the research associated with the detection, prevention and cure of the Alzheimer's disease. The patient or the family of the patient are not interested about the research article rather they search the documents that give the instant result about the cure or treatment of the disease. The research article described in [4] form LLM and Knowledge graph to answer few questions on Alzheimer's disease. The advancement in the area of Generative AI changes the way users interact with the system. The use of Large Language Model (LLM) [5] helps us to get curated answer from any GenAI based system. However LLM suffers from many problems such has hallucination, security issues. Henceforth Retrieval Augmented Generation (RAG) [6] is evolved to avoid the previously mentioned problems. Using the

RAG at backend AI chatbots are developed that use NLP [7] and Text mining [8] for human like conversation. These chatbots are improved with cognitive skills such that it able to engage clinicians and patients to discuss about patients' health conditions to come up with the dimension towards diagnosis and treatment. However concerns are there about the accuracy of the results that we get from the chatbot. The article in [9] discusses about the associated benefits, limitation, and risks of GPT-4 as an AI Chatbot for Medicine. A chatbot consists of two main components: a general-purpose AI system and a chat interface. This article [9] used GPT-4 (Generative Pretrained Transformer 4) with a chat interface. In the area of medical science the scope to have an error should be minimized as any error could be as fatal as death of a person. Hence precautions need to be taken to make sure that the result is correct so that based on that right answers are given to the user. These answers are the decisive factors about the treatment. Any research in the area of medical science based on Generative AI must focus on accurate answer retrieval. Moreover prompt engineering [10] can be incorporated to fine tune the answer.

3   Objectives

This research work aims to build a Retrieval Augmented Generation System for AD question-answering aims to provide accurate, contextually relevant, and user-friendly answers to queries related to Alzheimer's Disease. By leveraging cutting-edge generative AI and prompt engineering, the system will empower healthcare professionals, caregivers, and patients with timely access to valuable information, ultimately improving the understanding and management of AD. This system will enhance user experience and facilitate exploration of AD-related topics through the generation of related queries. The objective of this research work is pointed below:

   i. Providing accurate and up-to-date information using RAG.

   ii. Reducing hallucination using RAG and prompt engineering.

   iii. Storing the text data in the form of vector database by converting the text to vector and then applying chunking and embedding.

   iv. Related question generation by understanding the pattern of the questions from the user using prompt engineering.

4   Related Terminologies

4.a. Large Language Models(LLMs): Large language models (LLMs) are a type of foundational model trained on vast datasets, enabling them to understand and generate natural language as well as other content types to perform various tasks. These models utilize deep learning techniques and extensive textual data. Typically based on a transformer architecture, such as the generative pre-trained transformer (GPT), they excel at processing sequential data like text input. LLMs consist of multiple neural network layers, each with parameters that can be adjusted during training. An additional layer, known as the attention mechanism, enhances their capability by focusing on specific parts of the data.

During training, LLMs learn to predict the next word in a sentence by considering the context provided by the preceding words. They achieve this by assigning a probability score to the recurrence of tokenized words—these tokens are smaller sequences of characters. These tokens are then converted into embedding, which are numerical representations of the context. This training involves using an extensive corpus of text, spanning billions of pages, enabling the model to learn grammar, semantics, and conceptual relationships through zero-shot and self-supervised learning. Once trained, LLMs can generate text by predicting the next word based on the input they receive, utilizing the patterns and knowledge they have acquired. This results in coherent and contextually relevant language generation, useful for various natural language understanding (NLU) and content creation tasks.

Model performance can be enhanced through techniques like prompt engineering, prompt-tuning, and fine-tuning. Additionally, reinforcement learning with human feedback (RLHF) is used to mitigate biases, hateful speech, and "hallucinations" (factually incorrect answers) that can arise from training on vast amounts of unstructured data. Ensuring that enterprise-grade LLMs are reliable and safe for use is crucial to avoid potential liabilities and protect organizational reputations.

4.b. Vector Database: A vector database is designed to store, manage and index massive quantities of high-dimensional vector data efficiently. In contrast to traditional databases that handle tabular or document-based data, vector databases are optimized for managing spatial information represented as geometric shapes such as lines and polygons. Vector databases organize spatial data using a vector data model, which represents geometric objects as collections of vertices and edges. Each object is defined by its geometry (shape and location) and may also include additional attributes such as metadata or descriptive information.

Vector databases use various indexing techniques to enable faster searching. Vector indexing along with distance-calculating algorithms such as nearest neighbour search, are particularly helpful with searching for relevant results across millions if not billions of data points, with optimized performance.

Vector databases find applications in various domains, including urban planning, transportation management, environmental modelling, and asset tracking. They are particularly well-suited for scenarios that require complex spatial queries and advanced spatial analysis capabilities.

4.c. Prompt Engineering: Prompt engineering refers to the process of designing and crafting prompts or inputs to language models in order to elicit desired responses or behaviours. This technique has gained significant attention in the field of natural language processing (NLP), particularly with the rise of large language models (LLMs) such as OpenAI's GPT (Generative Pre-trained Transformer) series and Google's BERT (Bidirectional Encoder Representations from Transformers).

At its core, prompt engineering aims to shape the behaviour of language models by providing them with relevant context and instructions. This context could include keywords, phrases, or specific cues tailored to the task at hand. By carefully designing the input, prompt engineers aim to elicit responses from the language model that align with the desired task objectives.

Moreover, prompt engineering often involves iterative experimentation and evaluation. Engineers may test different variations of prompts, fine-tune language models on task-specific data, and analyse the quality of generated outputs. This iterative approach allows prompt engineers to refine their prompts, improve the performance of language models, and address any shortcomings or biases.

4.d. Retrieval Augmented Generation (RAG): Retrieval-Augmented Generation (RAG) is a novel approach in natural language processing (NLP) that integrates both retrieval-based and generation-based models. Unlike traditional generative models, RAG incorporates information retrieval techniques to retrieve relevant context from a large corpus of text data before generating a response. This retrieved context serves as input to the generation component, which then produces a response informed by the retrieved information. RAG models are trained using supervised learning and reinforcement learning techniques to optimize both the retrieval and generation components. They find applications in various NLP tasks such as conversational agents, question answering, summarization, and content generation, offering the advantage of producing more informative and contextually relevant responses compared to purely generative models. However, RAG also faces challenges such as scalability of the retrieval component, integration of retrieved context with the generation process, and potential biases in the retrieved data. Despite these challenges, RAG represents a promising approach in NLP and is expected to play a significant role in advancing the field.

5 Proposed Methodology

This research work utilizes a conversational AI system that combines advanced technologies to create an engaging and effective user experience. At the heart of the system is a sophisticated language model capable of understanding user inputs and generating contextually relevant responses. This allows the chatbot to hold natural and fluent conversations with users, making interactions feel human-like. The system uses memory management techniques to keep track of conversation history, which helps the chatbot provide more personalized and consistent answers over time.

To enhance the chatbot's performance and efficiency, the system leverages a specialized database designed for fast searching and data retrieval. This database stores data in a way that enables the system to quickly find the most similar information when a user asks a question. By efficiently matching user queries with relevant data, the system can deliver comprehensive and accurate responses to the user.

In this research work, the process of generating a question-answer (Q&A) model revolves around effectively managing and searching through a collection of text documents that have been split into smaller chunks and stored as embedding in a vector database. This system can be thought of as an advanced information retrieval and question-answering framework that leverages natural language processing and machine learning techniques. First, the text documents are processed and divided into manageable chunks. These chunks are then embedded using a pre-trained model, which transforms the text into high-dimensional vectors. By storing these vectors in a vector database we can efficiently retrieve and search for relevant information in response to user queries.

When a user submits a query, the query is refined to enhance its relevance and accuracy. This refinement may include pre-processing steps such as removing stop words, stemming, or lemmatization. Once the query is ready, it is embedded using the same model used for the text chunks. The query vector is then used to search the vector database to find the most similar or relevant text chunks. Once relevant chunks are identified, they are sent back to the user, along with the original query context, to provide a coherent and helpful response. This process can enhance the quality and speed of information retrieval, making it easier for users to access the precise information they need from a large collection of text data Fig. 1.

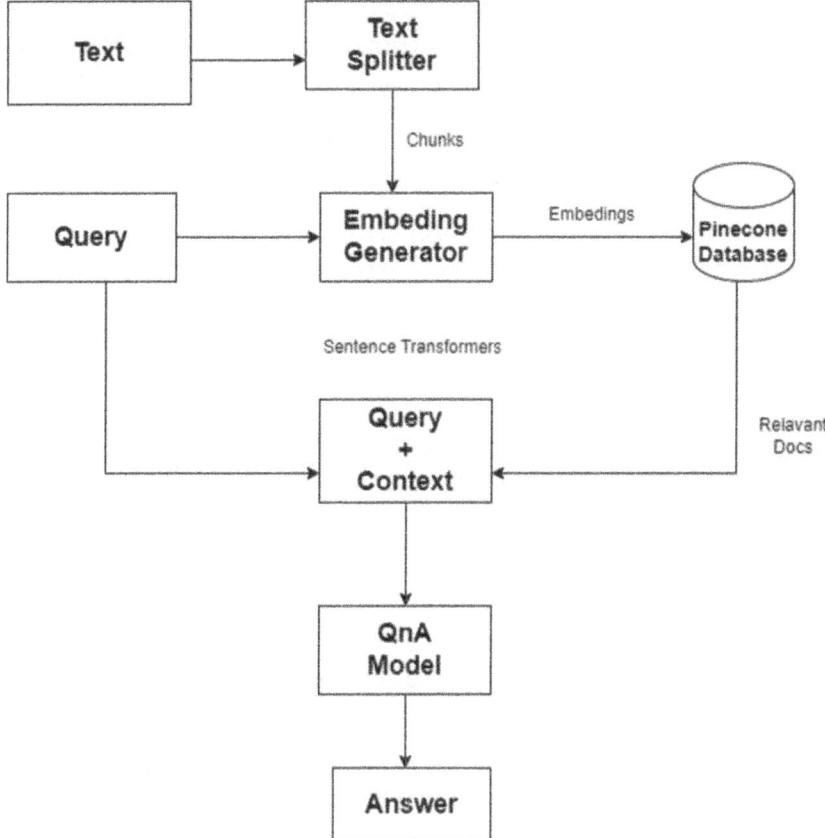

Fig. 1. Process Diagram of this RAG system

5.1 Customizing Prompt Engineering

In this research work we have used prompt engineering in three different places, i.e. Query Refiner, Related Questions Generator and System message template.

5.1. A) Query Refiner:

The Query Refiner method is designed to enhance the user's query by refining it for clarity, grammar, and spelling. The goal is to transform the initial user query into a well-formed question that can effectively retrieve relevant information from the knowledge base. The prompt for the query refiner is given below.

Prompt: "Given the following user query and conversation log, formulate a question that would be the most relevant to provide the user with an answer from a knowledge base.\n\nCONVERSATION LOG: \n{conversation}\n\nQuery: {query}\n\nRefined Query:"

The provided prompt guides this process, instructing the model to generate a refined question based on the user's original query and chat history. By ensuring that the refined question is free of grammatical and spelling errors, the prompt aims to improve the precision of the information retrieval process. The inclusion of an example in the prompt serves as a helpful guide for the language model, demonstrating the expected output format and style for the refined question. Overall, this approach enhances the quality and relevance of the responses provided to the user.

5.1. B) Related Questions Generator:

The Related Questions Generator is an essential part of providing additional context and depth to the user's query. This method generates three questions related to the user's initial query, offering a broader understanding of the topic and potentially enriching the user's experience with the knowledge base. The prompt for the related question generator is given below.

Prompt: "Provide three diverse, complete one-line questions related to '{query}' that delve into various aspects of the topic, ensuring they are straightforward and directly related. Avoid repeating the same content as the query. You need not to give any numbering of the question".

The prompt guides the model to generate three alternative questions based on the user's original query. These questions are intended to be straightforward, accessible, and directly related to the query. By offering these alternative questions, the model provides the user with a broader perspective on the topic and encourages further exploration. This enhances the overall interaction with the knowledge base and supports a richer, more informative experience for the user.

System Message Template:

The System Message Template in this work outlines how the AI model should handle user queries and provide responses based on the available context. This method sets expectations for the model's behaviour, ensuring it answers questions as accurately as possible using the provided information from the vector database. The prompt for the system message template is given below.

Prompt: "Answer the question as truthfully as possible using the provided context in a minimum of 200 words. Ensure the response is well-structured with proper spacing, and highlight important words in bold. If the answer includes multiple points, present the response in a clear point-by-point format with proper spacing between points. If the answer is not contained within the text below, say 'I do not know, because it is irrelevant to our context'".

The prompt for this method instructs the model to respond truthfully and transparently, indicating when it lacks sufficient information to provide a reliable answer. By explicitly stating "I DO NOT KNOW, because the data is not present in our Vector DB" when the necessary data is unavailable, the model fosters trust and clarity in the conversation. This approach helps manage user expectations and establishes a reliable standard for the AI's responses, enhancing the overall user experience and communication quality.

6 Case Study

- On the ChatUI page, the user inputs a prompt into the chatbot (Fig. 2). The prompt is refined (Fig. 2) and passed to the Pinecone vector database.
- The refined prompt generates a response and displays it (Fig. 2).
- Related questions based on the user's prompt are also generated (Fig. 3).
- If the query is not based on Alzheimer's, then our system will display "I do not know, because it is irrelevant to our context."(Fig. 4)

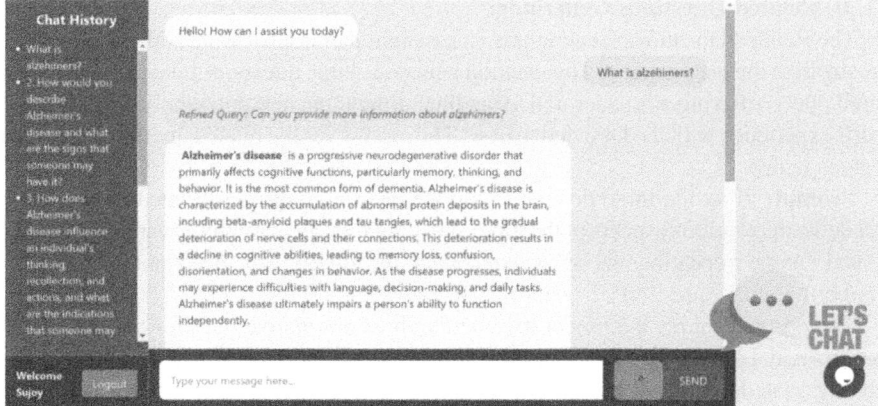

Fig. 2. Response of the query

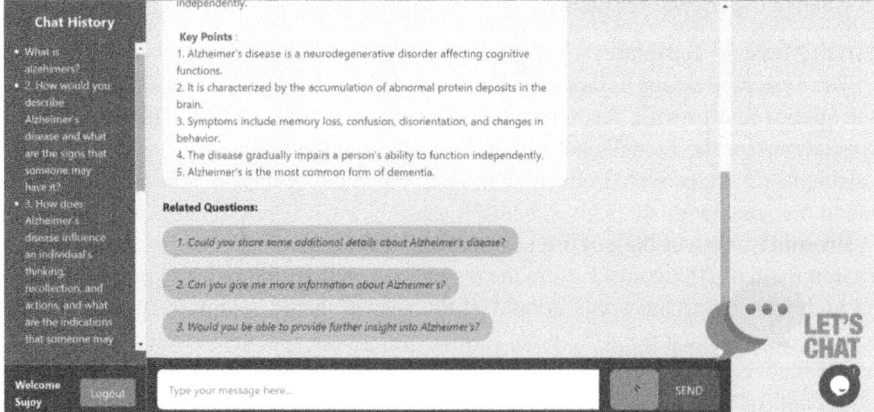

Fig. 3. Related questions of the given prompt

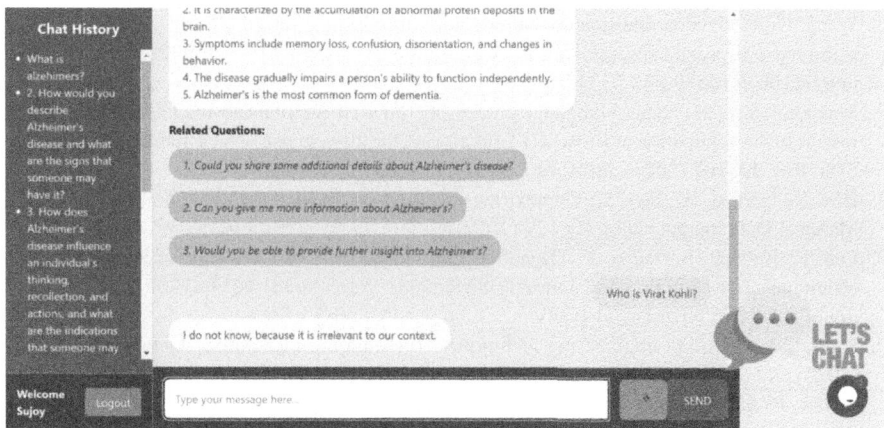

Fig. 4. If the query is irrelevant, the system replies this

7 Conclusion

This research work develops an interactive RAG based question answering system for Alzheimer's disease for doctor, patient, caregiver as well as the researchers in this domain. It uses the advanced techniques such vector database, vector embedding, text chunking, prompt engineering etc. Prompt engineer is used to perform query refinement, domain knowledge enhancement, response format. Moreover it is used to generate related question that help the user of the system to have better interaction. The system also able to reduces the hallucination using RAG and prompt engineering.

This system could be replicated to other domains by appropriately forming the RAG from the underlying domain. Moreover as new data are being gathered by the system through the interaction on Alzheimer's disease, this text corpus could be analysed further by using sentiment analysis [11] on the patient and well-wishers of patient for more effective information generation.

References

1. Saikia, P., Kalita, S.K.: Alzheimer disease detection using MRI: deep learning review. SN Comput. Sci. **5**, 507 (2024). https://doi.org/10.1007/s42979-024-02868-4
2. Alhyane, R., Kassimi El Bakkali, A., Bouroumi, A., Rémy, F., El Boustani, A.: "Detection of Alzheimer's Disease using a convolutional neural network". In: International Conference on Advanced Intelligent Systems for Sustainable Development. AI2SD (2022). https://doi.org/10.1007/978-3-031-35248-5_66
3. Li, F., Tran, L., Thung, K.H., Ji, S., Shen, D., Li, J.: A robust deep model for improved classification of AD/MCI Patients. IEEE J. Biomed. Health Inf. **19**(5) 1610−1616 (2015). https://doi.org/10.1109/JBHI.2015.2429556
4. Li, D., et al.: "DALK: Dynamic Co-Augmentation of LLMs and KG to answer Alzheimer's Disease Questions with Scientific Literature" (2024). [https://arxiv.org/pdf/2405.04819]
5. Wu, Y.: Large language model and text generation. In: Xu, H., Demner Fushman, D. (eds) Natural Language Processing in Biomedicine. Cognitive Informatics in Biomedicine and Healthcare (2024)

6. Wang, C., et al.: Potential for GPT technology to optimize future clinical decision-making using retrieval-augmented generation. Ann. Biomed. Eng. **52**, 1115−1118 (2024). https://doi.org/10.1007/s10439-023-03327-6

7. Masoumi, S., et al.: Natural language processing (NLP) to facilitate abstract review in medical research: the application of BioBERT to exploring the 20-year use of NLP in medical research. Syst. Rev. **13**, 107 (2024). https://doi.org/10.1186/s13643-024-02470-y

8. Roy, S., Cortesi, A., Sen, S.: Context-aware OLAP for textual data warehouses. Int. J. Inf. Manage. Data Insights **2**(2), 100129 (2022)

9. Lee, P., Bubeck, S., Petro, J.: "Benefits, limits, and risks of GPT-4 as an AI Chatbot for Medicine" The New England Journal of Medicine Vol. 388(13) 1233−1239 (2023) https://doi.org/10.1056/NEJMsr2214184v

10. Kansal, A.: "Prompt engineering techniques. In: Building Generative AI-Powered Apps." (2024). doi.org/https://doi.org/10.1007/979-8-8688-0205-8_8

11. Ghosh, P., Samanta, O., Goto, T., Sen, S.: Sales forecasting of overrated products: fine tuning of customer's rating by integrating sentiment analysis. IEEE Access **12**, 69578−69592 (2024)https://doi.org/10.1109/ACCESS.2024.3402133

Assessing the Impact of Prompt Strategies on Text Summarization with Large Language Models

Aytuğ Onan[1(✉)] and Hesham Alhumyani[2]

[1] Department of Computer Engineering, Faculty of Engineering and Architecture, İzmir Katip Çelebi University, 35620 İzmir, Turkey
aytug.onan@ikcu.edu.tr
[2] Department of Computer Engineering, College of Computers and Information Technology, Taif University, P.O. Box 11099 Taif 21944, Saudi Arabia
h.alhumyani@tu.edu.sa

Abstract. The advent of large language models (LLMs) has significantly advanced the field of text summarization, enabling the generation of coherent and contextually accurate summaries. This paper introduces a comprehensive framework for evaluating the performance of state-of-the-art LLMs in text summarization, with a particular focus on the impact of various prompt strategies, including zero-shot, one-shot, and few-shot learning. Our framework systematically examines how these prompting techniques influence summarization quality across diverse datasets, namely CNN/Daily Mail, XSum, TAC08, and TAC09. To provide a robust evaluation, we employ a range of intrinsic metrics such as ROUGE, BLEU, and BERTScore. These metrics allow us to quantify the quality of the generated summaries in terms of precision, recall, and semantic similarity. We evaluated three prominent LLMs: GPT-3, GPT-4, and LLaMA, each configured to optimize summarization performance under different prompting strategies. Our results reveal significant variations in performance depending on the chosen prompting strategy, highlighting the strengths and limitations of each approach. Furthermore, this study provides insights into the optimal conditions for employing different prompt strategies, offering practical guidelines for researchers and practitioners aiming to leverage LLMs for text summarization tasks. By delivering a thorough comparative analysis, we contribute to the understanding of how to maximize the potential of LLMs in generating high-quality summaries, ultimately advancing the field of natural language processing.

Keywords: Large Language Models · Text Summarization · Prompt Strategies · Zero-shot Learning · One-shot Learning · Few-shot Learning · ROUGE · BLEU · BERTScore

1 Introduction

The field of natural language processing (NLP) has experienced remarkable advancements with the advent of large language models (LLMs), such as GPT-3,

G. Hu et al. (Eds.): CAINE 2024, CCIS 2242, pp. 41–55, 2025.
https://doi.org/10.1007/978-3-031-76273-4_4

GPT-4, and LLaMA [1]. These models have demonstrated an impressive capability to generate coherent and contextually relevant text, significantly impacting various NLP tasks, including text summarization [2]. Text summarization, the process of condensing a text document to its essential information while retaining key points and overall meaning, is a crucial task in information retrieval and content management [3].

Recent advancements in LLMs have led to significant improvements in the quality of generated summaries [4]. However, the performance of these models is highly dependent on the strategies used to prompt them [5]. Prompting strategies, such as zero-shot, one-shot, and few-shot learning, play a critical role in guiding LLMs to produce high-quality summaries. Zero-shot learning involves generating summaries without any task-specific training examples, relying solely on the model's pre-trained knowledge. One-shot learning provides a single example to guide the summarization task, while few-shot learning involves a small number of examples to fine-tune the model's performance [6].

Despite the promising capabilities of LLMs, there remains a need for a systematic evaluation framework to assess the impact of different prompt strategies on summarization quality. This paper introduces a comprehensive framework for evaluating state-of-the-art LLMs in the context of text summarization. Our framework explores the effects of zero-shot, one-shot, and few-shot learning strategies on summarization performance across a diverse set of datasets, including CNN/Daily Mail, XSum, TAC08, and TAC09.

To ensure a robust evaluation, we employ a range of intrinsic metrics such as ROUGE, BLEU, and BERTScore. ROUGE (Recall-Oriented Understudy for Gisting Evaluation) measures the overlap of n-grams between the generated summary and reference summary, providing insights into the recall and precision of the summarization. BLEU (Bilingual Evaluation Understudy) focuses on the precision of n-grams, assessing how closely the generated summary matches the reference summary. BERTScore leverages pre-trained BERT embeddings to evaluate the semantic similarity between the generated and reference summaries, offering a more nuanced understanding of summary quality.

Our study aims to provide a comprehensive analysis of how different prompt strategies influence the performance of LLMs in text summarization. By examining the strengths and limitations of each approach, we offer practical guidelines for researchers and practitioners looking to leverage LLMs for summarization tasks. Additionally, our findings contribute to the broader understanding of LLM capabilities and their optimal use in various NLP applications.

The remainder of this paper is structured as follows: Section 2 reviews related work in the field of text summarization and LLM evaluation. Section 3 describes our framework, including the datasets and metrics used. Section 4 presents the experimental setup and results. Finally, Sect. 5 concludes the paper and discusses potential directions for future research.

2   Related Work

Text summarization is a well-established field in natural language processing (NLP) aimed at distilling the most informative parts of a text into a condensed form [7]. Traditionally, summarization methods have been categorized into extractive and abstractive approaches [8]. Extractive summarization techniques primarily focus on selecting key sentences from the original text, relying on statistical measures and heuristic methods. Early models utilized features like term frequency, key phrases, and sentence location to identify significant sentences [9].

With the rapid growth of digital information, the demand for more sophisticated summarization tools has increased. The introduction of machine learning has brought new approaches to both extractive and abstractive summarization. Unsupervised learning algorithms, such as Latent Semantic Analysis (LSA) and clustering-based methods, have been employed to enhance summary relevance and coherence without requiring extensive labeled data [10].

In recent years, the field has witnessed a paradigm shift with the advent of neural networks and deep learning models. The Transformer architecture, in particular, has revolutionized text summarization. Models like BERT and GPT have set new benchmarks, demonstrating superior performance in understanding and generating text by leveraging self-attention mechanisms and large-scale pre-training [11]. These models are capable of capturing complex sentence structures and contextual nuances, which are crucial for generating high-quality summaries.

The application of large language models (LLMs) in text summarization has opened new avenues for research, especially in the context of different prompt strategies. Zero-shot, one-shot, and few-shot learning represent significant advancements in leveraging LLMs without extensive task-specific training. Zero-shot learning generates summaries based on the model's pre-trained knowledge, while one-shot and few-shot learning utilize one or a few examples to guide the summarization process.

Several studies have explored the use of LLMs for text summarization. For instance, [12] demonstrated the effectiveness of BART, a denoising autoencoder for pre-training sequence-to-sequence models, in various summarization tasks. Similarly, [13] introduced PEGASUS, which achieved state-of-the-art results on summarization benchmarks by pre-training on a self-supervised objective tailored for abstractive summarization. However, these studies primarily focus on model architecture and training techniques rather than the impact of different prompt strategies.

Despite these advancements, evaluating the performance of LLMs across different prompting techniques remains an underexplored area. There is a need for a systematic framework to assess how zero-shot, one-shot, and few-shot learning strategies influence summarization quality across diverse datasets. This paper addresses this gap by introducing a comprehensive evaluation framework that utilizes intrinsic metrics such as ROUGE, BLEU, and BERTScore to measure the performance of LLMs in text summarization.

Our study builds on the existing body of work by providing a detailed analysis of prompt strategies and their effects on the quality of generated summaries. By leveraging a diverse set of datasets, including CNN/Daily Mail, XSum, TAC08, and TAC09, we offer insights into the optimal use of LLMs for summarization tasks. This research contributes to the broader understanding of how to effectively harness the potential of LLMs in NLP applications. While significant progress has been made in text summarization with the advent of LLMs and advanced prompt strategies, there is still a need for comprehensive evaluation frameworks to understand their true potential and limitations. Our study aims to fill this gap by systematically assessing the impact of different prompting techniques on summarization performance, providing valuable guidelines for future research and practical applications.

3   Methodology

3.1   Datasets

To ensure a robust and comprehensive evaluation of large language models (LLMs) for text summarization, we selected a diverse set of datasets. Each dataset represents different domains and types of text, providing a broad spectrum of summarization challenges. The chosen datasets are CNN/Daily Mail, XSum, TAC08, and TAC09. Below, we provide detailed descriptions of each dataset.

CNN/Daily Mail. The CNN/Daily Mail dataset is one of the most widely used benchmarks for text summarization. It comprises news articles paired with multi-sentence summaries. The dataset was created by extracting articles from the CNN and Daily Mail websites, where each article is associated with a summary written by human editors. The dataset contains approximately 287,000 training examples, 13,400 validation examples, and 11,500 test examples. Each example consists of a news article with an average length of 781 tokens and a summary with an average length of 56 tokens [14].

XSum. The Extreme Summarization (XSum) dataset is designed to encourage the development of abstractive summarization systems. Unlike the multi-sentence summaries in CNN/Daily Mail, XSum provides single-sentence summaries, typically created by the article's author, capturing the essence of the article in a concise manner. XSum consists of 204,045 articles extracted from the BBC website, split into 203,028 training examples, 11,332 validation examples, and 11,334 test examples. The average length of an article is 431 tokens, and the average length of a summary is 23 tokens [15].

TAC08. The Text Analysis Conference (TAC) 2008 dataset is a benchmark for summarizing complex documents, including news articles and technical reports.

Each document in the TAC08 dataset is paired with multiple reference summaries, providing a rich ground for evaluating summarization systems. The TAC08 dataset contains 48 document clusters, with each cluster containing 10 related documents. Each document has multiple human-written summaries, offering a robust evaluation setup [16].

TAC09. Similar to TAC08, the TAC 2009 dataset includes a variety of documents, making it another important benchmark for summarization tasks. The dataset aims to evaluate the consistency and reliability of summarization models across different document types and topics. TAC09 consists of 44 document clusters, each containing 10 documents. Each document cluster is paired with four reference summaries written by human annotators [16].

Dataset Preparation. For each dataset, we performed the following preprocessing steps to ensure compatibility with the LLMs:

- **Cleaning:** Removal of irrelevant metadata, advertisements, and non-textual elements.
- **Tokenization:** Splitting text into tokens suitable for input into the models.
- **Formatting:** Ensuring consistent input formats across all datasets.

These preprocessing steps are crucial for maintaining the quality and consistency of the input data, which directly impacts the performance of the summarization models.

Dataset Utilization. Each dataset was utilized to evaluate the performance of the LLMs under different prompt strategies. The variety of datasets allows us to assess how well the models generalize across different types of texts and summarization challenges. The diversity in content, structure, and summarization requirements across these datasets provides a comprehensive ground for evaluating the effectiveness of zero-shot, one-shot, and few-shot learning strategies.

3.2   Prompt Strategies

Prompting strategies are crucial for guiding large language models (LLMs) to perform specific tasks. In the context of text summarization, different prompt strategies can significantly impact the quality and coherence of the generated summaries. This section explores three primary prompting strategies: zero-shot learning, one-shot learning, and few-shot learning. Each strategy leverages varying amounts of task-specific information to fine-tune the model's output.

Zero-Shot Learning. Zero-shot learning refers to the ability of LLMs to perform a task without any task-specific training data. Instead, the model relies entirely on its pre-trained knowledge and the prompt provided at inference time.

For zero-shot summarization, we construct prompts that explicitly instruct the model to generate a summary of the given text. An example prompt might be:

"Summarize the following article: [Article Text]"

The model generates a summary based on its understanding of the instruction and the provided text.

Zero-shot learning demonstrates the inherent capabilities of LLMs to generalize and understand tasks without additional training. It is particularly useful in scenarios where task-specific training data is scarce or unavailable. The primary challenge with zero-shot learning is that the generated summaries may lack specificity and relevance due to the absence of task-specific fine-tuning. The model's reliance on pre-trained knowledge can result in summaries that are less tailored to the nuances of the input text.

One-Shot Learning. One-shot learning involves providing the model with a single example to guide the summarization task. This strategy leverages minimal task-specific data to improve the model's performance.

In one-shot summarization, we include a single example of a summarized text in the prompt. An example prompt might be:

"Here is an example of a summary:
Article: [Example Article Text]
Summary: [Example Summary]
Now, summarize the following article: [Article Text]"

This prompt provides the model with a clear example of the expected output, which it can use as a reference to generate the new summary.

One-shot learning helps the model to better understand the task requirements by providing a concrete example. This can lead to more relevant and accurate summaries compared to zero-shot learning. The effectiveness of one-shot learning depends heavily on the quality and relevance of the single example provided. An inappropriate or poorly constructed example can misguide the model, resulting in suboptimal summaries.

Few-Shot Learning. Few-shot learning involves providing the model with a small number of examples to guide the summarization task. This strategy uses limited task-specific data to significantly enhance the model's performance.

For few-shot summarization, we construct prompts that include several examples of summarized texts. An example prompt might be:

"Here are some examples of summaries:
Article 1: [Example Article Text 1]
Summary 1: [Example Summary 1]
Article 2: [Example Article Text 2]
Summary 2: [Example Summary 2]
Now, summarize the following article: [Article Text]"

The multiple examples provide the model with a broader understanding of the task, helping it to generate more accurate and coherent summaries.

Few-shot learning significantly improves the model's ability to generate high-quality summaries by leveraging multiple examples. It combines the benefits of task-specific fine-tuning with the flexibility of minimal data requirements. The main challenge with few-shot learning is the selection and construction of examples. The examples need to be representative of the variety of texts the model will encounter, ensuring that the model can generalize effectively from the few examples provided.

3.3   Evaluation Metrics

Evaluating the performance of large language models (LLMs) in text summarization tasks requires robust and reliable metrics. In this study, we employ a range of intrinsic evaluation metrics to assess the quality of the generated summaries. These metrics include ROUGE, BLEU, and BERTScore, each offering unique insights into different aspects of summarization performance. This section provides a detailed description of these metrics, their calculation methods, and their relevance to text summarization.

ROUGE. ROUGE (Recall-Oriented Understudy for Gisting Evaluation) is a set of metrics commonly used to evaluate automatic summarization and machine translation. ROUGE measures the overlap between the generated summary and reference summaries based on n-grams, word sequences, and word pairs. ROUGE-N measures the n-gram overlap between the generated summary and reference summaries. It is defined as [17]:

$$ROUGE\text{-}N = \frac{\sum_{\text{ref} \in \text{Refs}} \sum_{\text{gram}_n \in \text{ref}} \min(\text{Count}(\text{gram}_n), \text{Count}(\text{gram}_n))}{\sum_{\text{ref} \in \text{Refs}} \sum_{\text{gram}_n \in \text{ref}} \text{Count}(\text{gram}_n)} \tag{1}$$

where $\text{Count}(\text{gram}_n)$ is the number of times the n-gram appears in the generated summary and reference summaries. Commonly used n-grams are unigrams (ROUGE-1) and bigrams (ROUGE-2).

ROUGE-L measures the longest common subsequence (LCS) between the generated summary and reference summaries. The LCS captures the longest sequence of words that appear in both summaries in the same order. ROUGE-L is defined as:

$$ROUGE\text{-}L = \frac{LCS(\text{summary}, \text{reference})}{\text{Length}(\text{reference})} \tag{2}$$

ROUGE-W is a weighted LCS-based measure that assigns different weights to different parts of the LCS, giving more importance to consecutive matches. This helps in capturing the fluency and coherence of the generated summary. ROUGE metrics are widely used due to their simplicity and effectiveness in measuring the overlap between generated and reference summaries. They provide a straightforward way to evaluate precision, recall, and F1-score, making them useful for comparing different summarization models and strategies.

BLEU. BLEU (Bilingual Evaluation Understudy) is a precision-oriented metric initially developed for machine translation but also used in text summarization. BLEU measures how many words or phrases in the generated summary match those in the reference summary. BLEU calculates the geometric mean of modified precision scores for different n-grams (typically up to 4-grams) and applies a brevity penalty to penalize short summaries. It is defined as [18] :

$$\text{BLEU} = \text{BP} \cdot \exp\left(\sum_{n=1}^{N} w_n \log p_n\right) \tag{3}$$

where p_n is the precision for n-grams, w_n is the weight assigned to each n-gram, and BP is the brevity penalty.

The brevity penalty (BP) ensures that the generated summary is not too short compared to the reference summary. It is calculated as [18]

$$\text{BP} = \begin{cases} 1 & \text{if } c > r \\ \exp(1 - \frac{r}{c}) & \text{if } c \leq r \end{cases} \tag{4}$$

where c is the length of the generated summary and r is the length of the reference summary. BLEU focuses on precision, making it effective in evaluating the accuracy of word choices in the generated summary. However, it may not fully capture semantic meaning and coherence, which is why it is often used alongside other metrics.

BERTScore. BERTScore is a recent metric that leverages pre-trained BERT embeddings to evaluate the semantic similarity between the generated summary and reference summaries. It addresses some of the limitations of n-gram-based metrics by focusing on contextual and semantic similarity. BERTScore calculates the cosine similarity between the BERT embeddings of tokens in the generated and reference summaries. The metric computes precision, recall, and F1-score based on these similarities [19]:

$$\text{BERTScore Precision} = \frac{1}{|S|} \sum_{x \in S} \max_{y \in R} \text{cosine}(E(x), E(y)) \tag{5}$$

$$\text{BERTScore Recall} = \frac{1}{|R|} \sum_{y \in R} \max_{x \in S} \text{cosine}(E(y), E(x)) \tag{6}$$

$$\text{BERTScore F1} = 2 \cdot \frac{\text{BERTScore Precision} \cdot \text{BERTScore Recall}}{\text{BERTScore Precision} + \text{BERTScore Recall}} \tag{7}$$

where $E(x)$ and $E(y)$ are the BERT embeddings of tokens x and y in the generated summary S and reference summary R, respectively.

BERTScore is particularly useful for capturing the semantic similarity between summaries, providing a more nuanced evaluation of the generated text's quality. It is less sensitive to exact word matching and more focused on overall meaning and coherence, making it an excellent complement to traditional n-gram-based metrics.

3.4   Large Language Models

For this study, we employed several state-of-the-art LLMs known for their effectiveness in natural language understanding and generation tasks. The primary models evaluated include GPT-3, GPT-4, and LLaMA, each configured to optimize summarization performance.

GPT-3. GPT-3 (Generative Pre-trained Transformer 3) is a transformer-based model with 175 billion parameters, pre-trained on a diverse dataset encompassing a significant portion of the internet. For our experiments, we used the largest available version, GPT-3 Davinci, which offers the highest capacity for understanding and generating text [20].

GPT-4. GPT-4 (Generative Pre-trained Transformer 4) is an advanced version of the GPT series, featuring enhancements in terms of architecture and training data, allowing it to generate more accurate and contextually relevant summaries. It retains a similarly large parameter count and has been fine-tuned for various NLP tasks, including summarization [21].

LLaMA. LLaMA (Large Language Model Meta AI) is another state-of-the-art transformer-based model known for its efficiency and effectiveness in text generation tasks. It is pre-trained on a diverse dataset and fine-tuned for specific applications, including text summarization. LLaMA's architecture and training regimen enable it to produce high-quality summaries with improved coherence and relevance [22].

4   Results and Discussion

In this section, we present a detailed analysis of the performance of three large language models (LLMs) - GPT-3, GPT-4, and LLaMA - in text summarization tasks. In Tables 1, 2, and 3, the empirical results for GPT-3, GPT-4, and LLaMa have been presented, respectively. The analysis is based on their performance across different datasets (CNN/Daily Mail, XSum, TAC08, and TAC09) and prompt strategies (zero-shot, one-shot, and few-shot learning). The evaluation metrics used include ROUGE-1, ROUGE-2, BLEU, and BERTScore. The results demonstrate that GPT-4 consistently outperforms both GPT-3 and LLaMA across all datasets and evaluation metrics. This highlights GPT-4's advanced

Table 1. GPT-3 Results

Prompt Strategy	Dataset	ROUGE-1	ROUGE-2	BLEU	BERTScore
Zero-shot	CNN/Daily Mail	32.0	13.7	19.4	84.6
One-shot	CNN/Daily Mail	36.2	16.4	22.8	86.3
Few-shot	CNN/Daily Mail	39.8	19.0	25.1	88.2
Zero-shot	XSum	27.3	8.9	16.7	81.3
One-shot	XSum	30.6	11.2	18.5	83.0
Few-shot	XSum	34.4	13.4	21.2	85.1
Zero-shot	TAC08	33.7	13.3	20.1	85.2
One-shot	TAC08	38.0	15.9	23.0	87.0
Few-shot	TAC08	41.6	17.8	25.6	88.8
Zero-shot	TAC09	33.2	13.0	19.8	84.8
One-shot	TAC09	37.4	15.6	22.6	86.5
Few-shot	TAC09	40.9	17.5	25.2	88.4

Table 2. GPT-4 Results

Prompt Strategy	Dataset	ROUGE-1	ROUGE-2	BLEU	BERTScore
Zero-shot	CNN/Daily Mail	34.5	15.2	21.4	85.6
One-shot	CNN/Daily Mail	38.7	17.9	24.8	87.3
Few-shot	CNN/Daily Mail	42.3	20.5	27.1	89.2
Zero-shot	XSum	29.8	10.4	18.7	82.3
One-shot	XSum	33.1	12.7	20.5	84.0
Few-shot	XSum	36.9	14.9	23.2	86.1
Zero-shot	TAC08	36.2	14.8	22.1	86.2
One-shot	TAC08	40.5	17.4	25.0	88.0
Few-shot	TAC08	44.1	19.3	27.6	89.8
Zero-shot	TAC09	35.7	14.5	21.8	85.8
One-shot	TAC09	39.9	17.1	24.6	87.5
Few-shot	TAC09	43.4	19.0	27.2	89.4

capabilities in generating accurate and contextually relevant summaries. LLaMA also shows strong performance, generally outperforming GPT-3 but not reaching the levels of GPT-4. The superior performance of GPT-4 can be attributed to its larger model size and more extensive training data, which enable it to better capture the nuances of the text and generate high-quality summaries. The analysis of performance by prompt strategy reveals that few-shot learning consistently yields the highest scores across all models and metrics. This demonstrates the significant advantage of providing multiple examples to guide the summarization process. One-shot learning also improves performance compared to zero-shot learning but does not match the effectiveness of few-shot learning.

Table 3. LLaMA Results

Prompt Strategy	Dataset	ROUGE-1	ROUGE-2	BLEU	BERTScore
Zero-shot	CNN/Daily Mail	33.0	14.2	20.2	85.1
One-shot	CNN/Daily Mail	37.2	16.9	23.6	86.8
Few-shot	CNN/Daily Mail	40.8	19.5	25.9	88.7
Zero-shot	XSum	28.3	9.4	17.5	81.8
One-shot	XSum	31.6	11.7	19.3	83.5
Few-shot	XSum	35.4	13.9	22.0	85.6
Zero-shot	TAC08	34.7	13.8	20.9	85.7
One-shot	TAC08	39.0	16.4	23.8	87.5
Few-shot	TAC08	42.6	18.3	26.4	89.3
Zero-shot	TAC09	34.2	13.5	20.6	85.3
One-shot	TAC09	38.4	16.1	23.4	87.0
Few-shot	TAC09	41.9	18.0	26.0	88.9

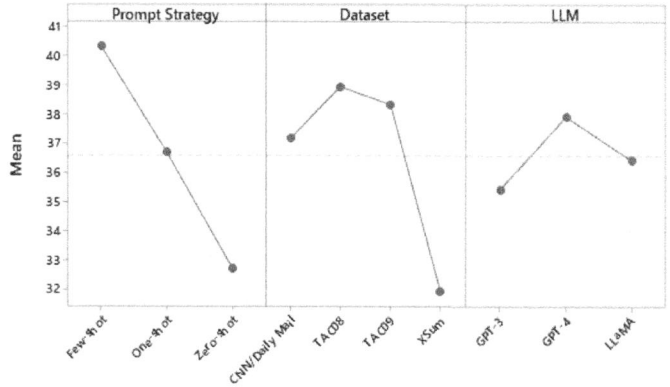

Fig. 1. ROUGE-1 scores

For GPT-3, the ROUGE-1 scores on the CNN/Daily Mail dataset increase from 32.0 in the zero-shot setting to 39.8 in the few-shot setting, highlighting the benefits of additional examples. Similarly, GPT-4 shows a marked improvement with few-shot learning, achieving ROUGE-1 scores of 42.3 compared to 34.5 in the zero-shot setting. LLaMA also demonstrates strong improvements with few-shot learning, achieving ROUGE-1 scores of 40.8 on the CNN/Daily Mail dataset compared to 33.0 in the zero-shot setting.

When examining performance across different datasets, GPT-4 achieves the highest scores, demonstrating its robustness and versatility in handling various

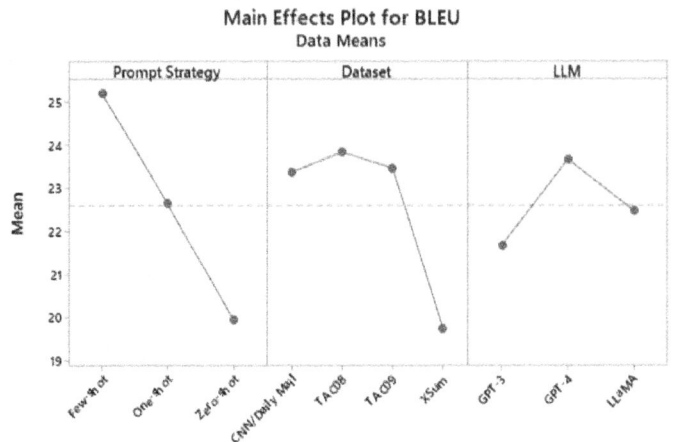

Fig. 2. BLEU scores

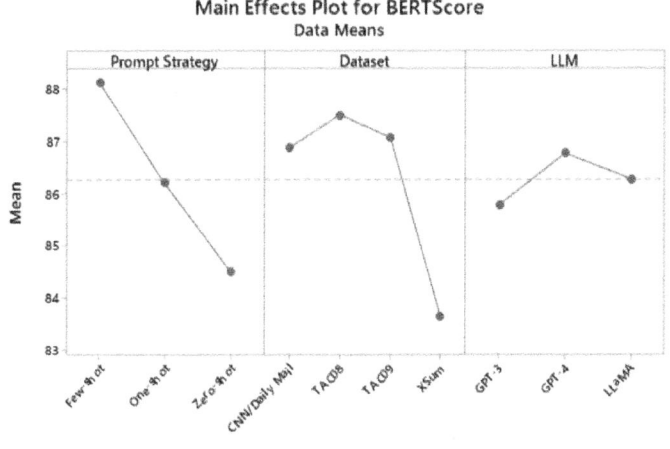

Fig. 3. BERT Scores

types of texts. For instance, on the TAC08 dataset, GPT-4 achieves ROUGE-1 scores of 44.1 in the few-shot setting, compared to 36.2 in the zero-shot setting. LLaMA shows strong performance particularly in the TAC08 and TAC09 datasets, which involve complex document summarization tasks. For example, LLaMA achieves ROUGE-1 scores of 42.6 on the TAC08 dataset in the few-shot setting. GPT-3, while effective, consistently scores lower than both LLaMA and GPT-4, highlighting the advancements made in the newer models.

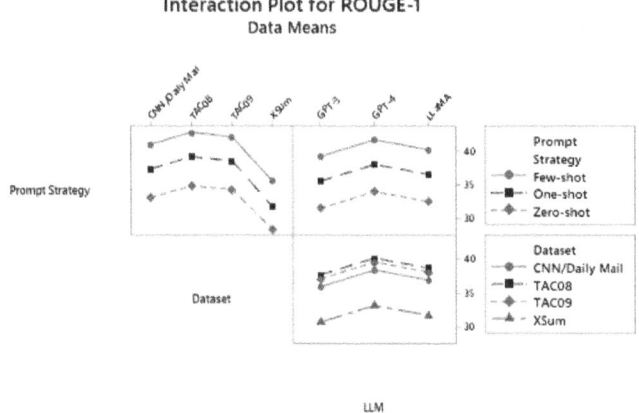

Fig. 4. Interaction Plot for ROUGE-1

In Figs. 1, 2, and 3, the main effect plots for ROUGE-1, BLUE, and BERTScore values, respectively. In addition, in Fig. 4, we present the interaction plot for ROUGE-1 score. The comparative analysis reveals several important insights. Few-shot learning is the most effective prompting strategy for all models, significantly improving the quality of generated summaries by leveraging multiple examples. This strategy allows the models to better understand the context and structure of the task, resulting in more accurate and coherent summaries. GPT-4's superior performance can be attributed to its larger model size and more extensive training data, enabling it to generate more contextually relevant and detailed summaries. The consistent outperformance across all datasets and metrics underscores the advancements in its architecture and training methodology. LLaMA offers a competitive alternative to GPT-4, particularly for complex summarization tasks. While it does not quite reach the performance levels of GPT-4, it consistently outperforms GPT-3, making it a viable option for high-quality summarization. GPT-3, while still a powerful model, shows the limitations of earlier LLMs compared to the latest advancements. The performance gap between GPT-3 and the newer models highlights the importance of continued model development and training on diverse and extensive datasets.

5   Conclusion

In this study, we presented a comprehensive framework for evaluating the performance of state-of-the-art large language models (LLMs) in text summarization tasks. Our framework systematically investigated the impact of different prompt strategies, including zero-shot, one-shot, and few-shot learning, across a diverse set of datasets: CNN/Daily Mail, XSum, TAC08, and TAC09. The evaluation employed intrinsic metrics such as ROUGE, BLEU, and BERTScore to assess the quality of the generated summaries.

The experimental results demonstrated that few-shot learning consistently outperforms zero-shot and one-shot learning across all datasets and metrics. This indicates the significant advantage of providing multiple examples to guide the model, enhancing its ability to generate accurate, coherent, and semantically rich summaries. One-shot learning also showed noticeable improvements over zero-shot learning, highlighting the value of even a single example in guiding the model's summarization task.

Our findings suggest several key implications for future research and practical applications. Firstly, optimizing the number and quality of examples in few-shot learning can further enhance summarization performance. Secondly, evaluating LLMs on additional datasets from various domains can provide insights into their generalization capabilities and limitations. Additionally, incorporating human evaluations alongside intrinsic metrics can offer a more holistic assessment of summary quality. Finally, exploring model interpretability can deepen our understanding of how LLMs utilize prompt examples, guiding the development of more effective prompting strategies.

This study underscores the potential of few-shot learning in advancing the performance of LLMs for text summarization. By leveraging task-specific examples, models can produce high-quality summaries that closely match human-written references, paving the way for more sophisticated and practical applications in various fields. Our framework and findings contribute to the broader understanding of LLM capabilities and their optimal use in text summarization, providing a solid foundation for future research and development in this area.

Acknowledgements. This research was supported by Taif University Researchers Supporting Project number TURSP-HC2024/13, Taif University, Saudi Arabia.

References

1. Chang, Y., et al.: A survey on evaluation of large language models. ACM Trans. Intell. Syst. Technol. **15**(3), 1–45 (2024)
2. Min, B., et al.: Recent advances in natural language processing via large pre-trained language models: a survey. ACM Comput. Surv. **56**(2), 1–40 (2023)
3. El-Kassas, W.S., Salama, C.R., Rafea, A.A., Mohamed, H.K.: Automatic text summarization: a comprehensive survey. Expert Syst. Appl. **165**, 113679 (2021)
4. Marvin, G., Hellen, N., Jjingo, D., Nakatumba-Nabende, J.: Prompt engineering in large language models. In: Jacob, I.J., Piramuthu, S., Falkowski-Gilski, P. (eds.) International Conference on Data Intelligence and Cognitive Informatics, pp. 387–402. Springer Nature, Singapore (2023). https://doi.org/10.1007/978-981-99-7962-2_30
5. Velásquez-Henao, J.D., Franco-Cardona, C.J., Cadavid-Higuita, L.: Prompt engineering: a methodology for optimizing interactions with AI-language models in the field of engineering. Dyna **90**(230), 9–17 (2023)
6. Liu, P., Yuan, W., Fu, J., Jiang, Z., Hayashi, H., Neubig, G.: Pre-train, prompt, and predict: a systematic survey of prompting methods in natural language processing. ACM Comput. Surv. **55**(9), 1–35 (2023)

7. Allahyari, M., et al.: Text summarization techniques: a brief survey. arXiv preprint arXiv:1707.02268 (2017)
8. Gambhir, M., Gupta, V.: Recent automatic text summarization techniques: a survey. Artif. Intell. Rev. **47**(1), 1–66 (2017)
9. Meena, S.M., Ramkumar, M.P., Asmitha, R.E., G SR, E.S.: Text summarization using text frequency ranking sentence prediction. In: 2020 4th International Conference on Computer, Communication
10. Abualigah, L., Bashabsheh, M.Q., Alabool, H., Shehab, M.: Text summarization: a brief review. In: Abd Elaziz, M., Al-qaness, M.A.A., Ewees, A.A., Dahou, A. (eds.) Recent Advances in NLP: The Case of Arabic Language. SCI, vol. 874, pp. 1–15. Springer, Cham (2020). https://doi.org/10.1007/978-3-030-34614-0_1
11. Liu, Y., Lapata, M.: Text summarization with pretrained encoders. arXiv preprint arXiv:1908.08345 (2019)
12. Gupta, A., Chugh, D., Anjum, Katarya, R.: Automated news summarization using transformers. In: Aurelia, S., Hiremath, S.S., Subramanian, K., Biswas, S.K. (eds.) Sustainable Advanced Computing: Select Proceedings of ICSAC 2021, pp. 249–259. Springer, Singapore (2022). https://doi.org/10.1007/978-981-16-9012-9_21
13. Zhang, J., Zhao, Y., Saleh, M., Liu, P.: PEGASUS: pre-training with extracted gap-sentences for abstractive summarization. In: International Conference on Machine Learning, pp. 11328–11339. PMLR (2020)
14. Li, J., Zhang, C., Chen, X., Cao, Y., Liao, P., Zhang, P.: Abstractive text summarization with multi-head attention. In: 2019 International Joint Conference on Neural Networks (IJCNN), pp. 1–8. IEEE (2019)
15. Hasan, T., et al.: XL-sum: large-scale multilingual abstractive summarization for 44 languages. arXiv preprint arXiv:2106.13822 (2021)
16. Nawrath, M., et al.: On the role of summary content units in text summarization evaluation. arXiv preprint arXiv:2404.01701 (2024)
17. Steinberger, J., Ježek, K.: Evaluation measures for text summarization. Computing and Informatics **28**(2), 251–275 (2009)
18. Suleiman, D., Awajan, A.: Deep learning based abstractive text summarization: approaches, datasets, evaluation measures, and challenges. Math. Probl. Eng. **2020**, 1–29 (2020)
19. Zhang, T., Kishore, V., Wu, F., Weinberger, K.Q., Artzi, Y.: BERTScore: evaluating text generation with BERT. arXiv preprint arXiv:1904.09675 (2019)
20. Dale, R.: GPT-3: What's it good for? Nat. Lang. Eng. **27**(1), 113–118 (2021)
21. Achiam, J., et al.: GPT-4 technical report. arXiv preprint arXiv:2303.08774 (2023)
22. Coda-Forno, J., Binz, M., Akata, Z., Botvinick, M., Wang, J., Schulz, E.: Meta-in-context learning in large language models. Adv. Neural. Inf. Process. Syst. **36**, 65189–65201 (2023)

Evaluation of the Effectiveness of Prompts and Generative AI Responses

Ajay Bandi[1(✉)] and Ruida Zeng[2]

[1] School of Computer Science and Information Systems, Northwest Missouri State University, 800 University Dr, Maryville, MO 64468, USA
AJAY@nwmissouri.edu
[2] Department of Computer Science, Brown University, Providence, RI 02912, USA

Abstract. Our paper proposes a comprehensive framework to evaluate the effectiveness of prompts and the corresponding responses generated by Generative Artificial Intelligence (GenAI) systems. To do so, our evaluation framework incorporates both objective metrics (accuracy, speed, relevancy, and format) and subjective metrics (coherence, tone, clarity, verbosity, and user satisfaction). A sample evaluation is performed on prompts send to Gemini and ChatGPT GenAI models. Additionally, our evaluation framework employs various feedback mechanisms, such as surveys, expert interviews, and automated reinforcement learning from human feedback (RLHF), to iteratively enhance the performance and reliability of GenAI models. By providing a holistic approach to evaluating and improving prompt-response effectiveness, our evaluation framework contributes to the development of more credible and user-friendly AI systems.

Keywords: Generative AI (GenAI) · Prompt Engineering · Evaluation Framework · Reinforcement Learning from Human Feedback (RLHF)

1 Introduction

Generative Artificial Intelligence (GenAI) represents a significant advancement in the field of AI in recent years, and is looking to further revolutionize various aspects of internet, technology, and even humanity [4].

Nevertheless, despite the consensus that Generative AI (GenAI) will play a pivotal role in shaping the future with incredible transformative potentials, there are many limitations in the current GenAI models [4]. A notable concern here is "hallucinations", where AI confidently gives out a factually incorrect response [29]. This is very concerning since the GenAI will also word the response to appear truthful and credible, and it rarely gives out a "Sorry I don't know..." response since it heavily prefers answers (regardless of correctness) to non-answers [29]. This is because at this time, no GenAI models actually understand the responses that it provide to the users, they just use probabilistic statistical model to select the correct words from the massive natural language database, and regurgitate those words back to the users [6,22]. In fact, researchers have coined the term "stochastic parrot" to criticize the phenomenon

G. Hu et al. (Eds.): CAINE 2024, CCIS 2242, pp. 56–69, 2025.
https://doi.org/10.1007/978-3-031-76273-4_5

of GenAI not genuinely understanding the underlying meaning of the words they read and write [6]. Sometimes, users are not satisfied with GenAI's responses due to a variety of reason, ranging from inaccurate information in the response, to the answer simply not being the one that user is looking for and want to see [16].

There are many ways to address this issue, such as continual improvement of the GenAI models by adding more training data, incorporating more advanced machine learning methods and techniques, and using better foundational large language models [5]. In addition, researchers have discovered that writing better prompts can often times result in better responses from the GenAI [9,32].

In order to improve the GenAI responses, our paper propose a holistic and comprehensive evaluation framework to evaluate the prompt-response effectiveness by measuring the performance of the GenAI responses to align with objective expectations and user experiences. While most of the framework focuses on the GenAI responses, it also considers metrics that affect both prompts and responses, such as semantic correctness, clarity, and verbosity. With proper and uniformed evaluations, methods such as reinforcement learning can then be used to enhance the overall effectiveness and reliability of the GenAI systems [19].

The remainder of the paper is organized as follows: Section 2 will contain works related to prompt engineering and motivation for the holistic evaluation framework; Sect. 3 will contain our proposed framework in detail, including objective metrics, subjective metrics, and feedback mechanisms; Sect. 4 will contain a sample evaluation of our proposed framework using Gemini and ChatGPT; Sect. 5 will contain our conclusion and future works.

2   Related Work

In recent years, prompt engineering has been an important topic for many academic researchers, and many work has been done on how writing better prompts can result in better responses from the GenAI, with most of the work focusing on the ChatGPT model specifically [9,11,30,32]. Other GenAI models such as Gemini were also explored for prompt engineering [26]. Overall, recent research has pointed out that prompt engineering has many applications such as in academic writing, medical education, and legal research where factually accurate responses are crucial [11,13,18,20,21]. However, most of the past work uses various different, sometimes arbitrary, evaluation to determine whether or not the GenAI responses are considered to be good or not, which is why our paper presents a holistic evaluation framework to assist with the evaluation of prompt-response effectiveness.

3   Proposed Framework

The proposed framework for the evaluation of effectiveness for prompts and their responses shown in Fig. 1 aims to systematically assess and enhance the quality of interactions between users and GenAI. It incorporates both objective and subjective metrics to provide a comprehensive evaluation. Objective metrics

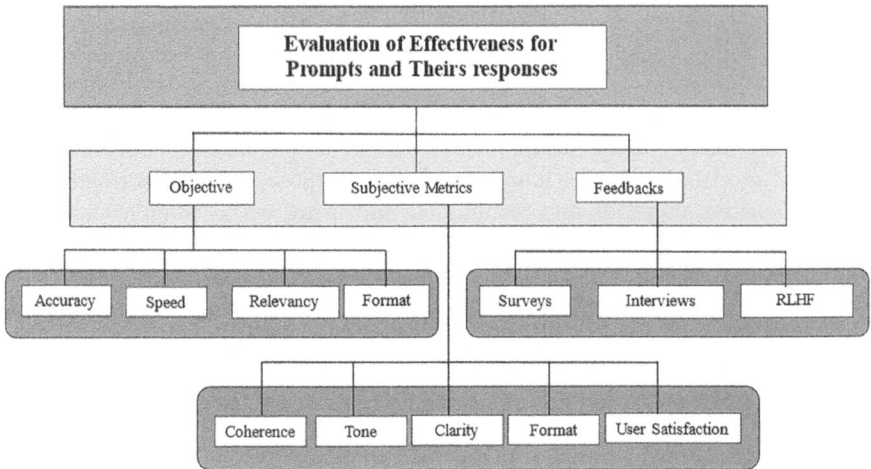

Fig. 1. Framework for Evaluating Prompt-Response Effectiveness

include accuracy, speed, relevancy, and format, ensuring that responses meet the standards. Subjective metrics focus on coherence, tone, clarify, verbosity, and user satisfaction, focusing more on the nuanced human experience of the interaction.

Additionally, we also introduce several feedback mechanisms, allowing for iterative improvements based both the objective and subjective metrics mentioned. This holistic approach ensures that the evaluation framework not only measures the responses both quantitatively and qualitatively with feedbacks that are designed for continuous improvement and enhancement of the GenAI.

3.1 Objective Metrics

Objective metrics provide quantifiable measures of performance, allowing for a clear and unbiased assessment of the GenAI system's capabilities. These metrics are essential for evaluating the technical aspects of the prompts and responses, ensuring that the system meets specific standards of correctness, quality, and efficiency. The key objective metrics here include accuracy, speed, relevancy, and format, as shown in Fig. 2.

Accuracy. Accuracy refers to both the factual correctness and the semantic correctness of the information provided.

It is crucial that the GenAI system delivers factually correct and precise answers that are not results of AI hallucinations to user queries [29]. Incorrect answers, especially incorrect answers delivered by the GenAI using confident languages, can be detrimental depending on the context and may cause legal and ethical concerns [21,29]. Semantic correctness is also important and ensures that the response is in the intended languages and not gibberish (unless specifically requested), and conveys the intended meaning accurately and logically.

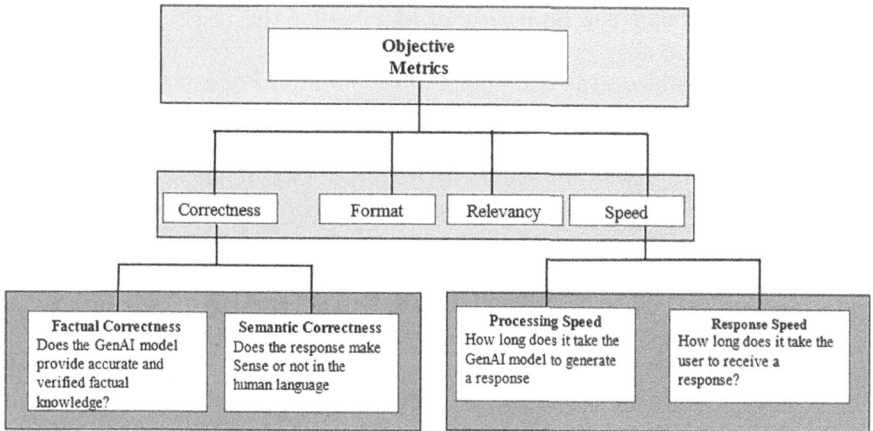

Fig. 2. Objective Metrics: Accuracy, Speed, Relevance, Format Assessment

It is worth nothing that factual incorrectness in the prompt itself might lead to hallucinations, and semantic incorrectness, such as stating the prompt in a language not supported by GenAI, would lead to GenAI not understanding the prompt at all [29].

Generally speaking, factual accuracy can be measured based on how closely the responses align with the established knowledge, data, or observable reality. This type of correctness is often verifiable through evidence, research, or documentation. For example, bogus law cases submitted to court by lawyers as part of their GenAI assisted legal research are clearly factually incorrect and inaccurate [21]. For AI generated source codes, this can be measured by whether or not the source codes correctly achieve the desired functionalities or produce the desired outputs when executed. For outputs such as images or voices, assessing the accuracy are a bit trickier. Semantic accuracy, on the other hands, can be measured based on whether or the generated languages use the correct grammar.

For AI generated source codes, semantic correctness can be measured by whether the source codes can compile successful in the given programming language. Unlike textual responses, visual and auditory outputs are inherently more subjective and difficult to measure their accuracy using objective metrics. For example, an image generated to match a prompt might be accurate in terms of content but the style and production may not meet user expectations.

Overall, accuracy is an extremely important metric and high accuracy ensures that users can trust the information provided by the system, which is essential for maintaining GenAI's credibility and user confidence in the responses [2].

Speed. The speed metric includes two major aspects: processing speed and response speed. Processing speed measures how long it takes the GenAI model to generate a response, and response speed measures how long it takes for the user to receive the response [1].

As a standalone metric, the higher the speed (both processing and response) the better. High speed GenAI can help speed up critical decision making and information retrieval in time sensitive situations for the users.

Relevancy. Relevancy can be measured by whether the responses answer the correct question, and not a completely unrelated question, or even a related question that does not address the main point of the prompt. For example, a question about std::array in C++ should not focus on C++ std::vector, even though they are similar constructs [25]. Another way relevancy can be defined would be how "up to date" the responses are, particularly in fields where knowledge is rapidly advancing, such as in science, technology, and medicine. An example of this would be the COVID-19 virus that caused a global pandemic a few years ago. Our knowledge and understanding about the virology of COVID-19 is now built upon a few years of solid worldwide medical research, and GenAI responses that align with information from contemporary sources is definitively more relevant than information from sources from 2021, even if both sources come from reliable medical professionals [8]. Needless to say, the more relevant the responses are with respect to the prompts, the more effective the responses are.

Format. Last but not least, we can measure whether or not the responses are delivered in the correct format. Some of the formats available include short sentences, paragraphs, bullet points, tables, equations, source code blocks, images, and even voices [1]. If the prompt asks the GenAI to generate an image, the GenAI's response should be in the image format, and not in the paragraph format describing the image textually. And if the prompt asks the GenAI to describe an algorithm that sorts a list, the GenAI's response should be in paragraph or bullet points format, rather than a code block that sorts a list using the built-in "sort()" method in Python [27].

Note that in the case where multiple response formats are acceptable, we can score the responses based on whether or not it is delivered in the optimal format. However, there could be cases where the "optimal format" is decided subjectively as opposed to objectively.

3.2 Subjective Metrics

Subjective metrics provide a qualitative assessment of the GenAI system's performance, focusing on the end user's experience aspects of human-computer interactions. These metrics capture the human elements of communication, such as how the responses are perceived and understood by users. It also evaluates whether the humans are satisfied with the interactions with the AI regardless of the accuracy or efficiency of the responses - although those factors can play a substantive part in the user satisfaction metric, different users might perceive and value them differently. By evaluating coherence, tone, clarity, verbosity, and user satisfaction, subjective metrics help ensure that the responses overall create a good experience for the users.

Coherence. Coherence refers to the logical sense and linguistic consistency of the response on its own without taking into other objective metrics. We can measure coherence based on the response's logical structure and whether or not it is easy to follow for the particular user. A coherent response would be a response that has clear connection between different parts and the overall message is cohesive. A response that would score poorly in the coherence category would be a response to abruptly jumps from one point to another with no transition, or presents random information throughout the response without mentioning how the information relate to the overall message that the response is trying to convey.

Tone. Tone is an important metric since an appropriate tone can make the responses seem more human-like, and GenAI should be able to adapt to different tones based on the context and the audience. Using deep learning techniques, GenAI models are able to detect user's emotions from the prompts and take that into consideration when generating responses [7]. If the prompt is just a casual question, then GenAI should use a more relaxed and friendly tone, and make the conversation more engaging. If the prompt is a professional inquiry, then GenAI should respond formally without using any slang. If the prompt addresses a sensitive topic, then GenAI should deliver the response with an empathetic and supportive tone, with lots of respect and consideration to the user's emotional feelings [31].

Clarity. Clarify refers to how understandable and clear the prompts and responses are. Clarify in prompts mean that the users should provide clear and specific instructions so that the GenAI understand exact what the users want [33]. With better prompts, the generated responses more accurate and relevant [33]. Clarity in responses means that GenAI should avoid using professional jargons or overly complicated technical terms unless they are essential to the responses. In cases where they are necessary, appropriate explanations and definitions should be provided to avoid confusions.

Verbosity. Verbosity refers to the conciseness of the response. Users generally have the options to specify the verbosity by explicitly requesting responses of a specific length or by asking the model to provide more or less information [1,9]. However, when unspecified, we want GenAI to provide useful information with sufficient details that answers the prompts fully with enough supporting evidence and reasoning, without omitting any important details, and without being overly verbose. Lengthy descriptions with unnecessary elaboration, excessive explanation for the obvious, or repetitions can cause frustration for the users, especially if they are purposefully looking for a concise and straightforward answer.

User Satisfaction. For our framework, "user satisfaction" serves as an umbrella term that encompasses all the other subtle subjective metrics and factors not already mentioned that contribute to the overall user experience and how pleased the users are with the prompt-response interaction. This can vary between different users and include things like preferences, where one user simply like the response more than another user, even when it is the exact same response to the exact same prompt. As mentioned before, sometimes even factors that are usually considered to be objective metrics can play a substantive part in the user satisfaction metric, since different users might perceive and value each objective metric differently.

3.3   Feedback

Feedback mechanisms are crucial for the continuous improvement and enhancement of the GenAI system [24,28]. In our framework, multiple feedback channels are employed to measure different aspects of user interaction and system performance to achieve comprehensive insights. This section outlines the key feedback mechanisms: surveys, interviews, and reinforcement learning from human feedback (RLHF).

Surveys. Surveys provided to the users after each response are a great way to collect instant feedback from the users about the interactions and measure the subjective metrics of the GenAI system. For example, the survey if the user like the responses or not (for user satisfaction metric), and also the user the rate the coherence, tone, clarity, verbosity of the responses from a scale of 1 to 10.

Interviews. Interviews with experts with knowledge in the fields can help measure the objective metrics of the GenAI responses. Some of the objective metric such as accuracy and relevancy may require academic or professional experts to verify how closely the responses align with the established knowledge, data, and observed reality, especially if the ground truth can not be easily determined by an average person. These kinds of interviews require that the experts involved are indeed experts and can be trusted to make a judgmental decision on the responses.

We can also ask the regular users about the subjective metrics in the interviews as well or even inquire users about potential areas for improvements in general.

RLHF. The results from both the surveys and the interviews can be converted into a "score" and fed into a model that uses reinforcement learning from human feedback (RLHF) techniques that use reward models to iteratively improve the prompt-response effectiveness overtime [14,17,24,28].

However, sometimes GenAI's surveys with long and excessive multiple choice questions can lead to survey fatigue and frustrate the user, causing the users to quickly select an inaccurate response and the users might spam click a random response that is the most convenient for them ergonomically, similar to how people might select January 1 as their birthday when registering an account online [15]. This will lead to the survey results being inaccurate and counterproductive for improving the prompt-response effectiveness.

Instead, we propose an automated feedback process with a chain of large language models (LLMs) and Non-LLMs that are RLHF for the user's usage

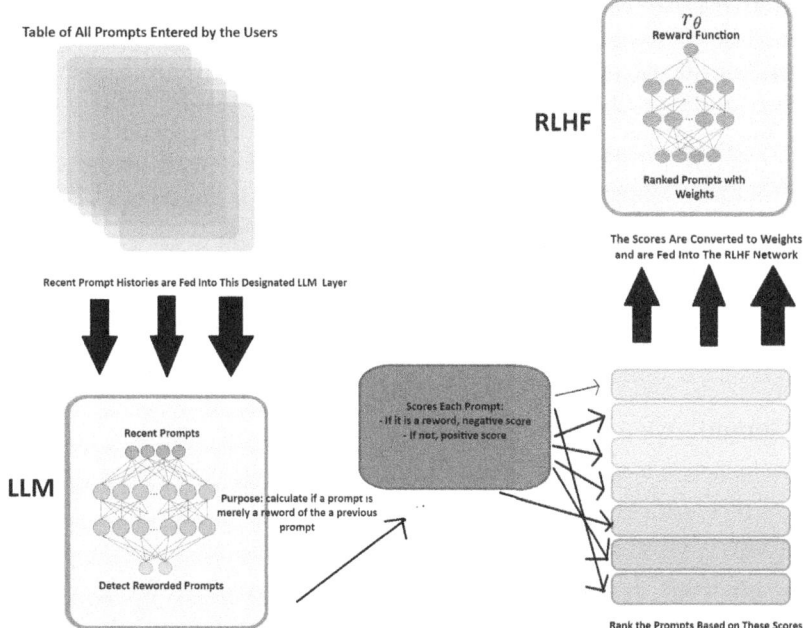

Fig. 3. Integrated Process of LLM with RLHF Utilizing User Feedback for Continuous Performance Enhancement

over time. One simple approach of doing this practically would be keeping a table of all the prompts entered by the users. Then, a layer of LLMs is used to calculate if a prompt is merely a reword of the a previous prompt by analyzing recent prompt history in the table. If so, this indicates the previous response corresponding to the previous prompt is a failed and unproductive response, and receives a negative score in the RLHF. Otherwise, it receives a positive score in the RLHF. Although this is done under the naive assumption that not rewording a prompt means that the response are "okay" and avoids the frustration of choice prompts. This method is shown in Fig. 3 [10].

Additionally, we can also use the same tools alongside a human sampling model to automate the scoring of expert-level prompts and responses by assessing how knowledgeable the users are with respect to the topic and complexity [3]. A secondary LLM can be used to categorize the questions by topic and complexity, then the human sampling framework is used to assess the user's expertise and historical performances. The GenAI can also measure the accuracy of user's judgment on information with established ground truth. If the user's decision on the accuracy of a response differs from the established ground truth, then the user gets a negative score for the topic and complexity in the human sampling RLHF, and vice versa. This way, we can determine which user's feedbacks are more valuable and should be associated with a higher confidence level score.

Further research could explore the use of more advanced RLHF to automate feedback mechanisms, eliminating the need for surveys and interviews.

4   Evaluation of Framework

In order to demonstrate an evaluation of the framework, we will be evaluating the GenAI responses to carefully designed prompts related to computer science for simplicity purpose, using our proposed framework. For our purpose, Gemini 1.0 and ChatGPT 3.5 will be used on the same sets of standardized prompts to ensure that the GenAI responses are comparable [12,23]. Relevant objective and subjective metrics will be collected and evaluated.

As a demonstration, we input the same prompt to both Gemini 1.0 and ChatGPT 3.5 for evaluation: **Please generate a short, concise, functioning Python code that demonstrates how recursion works.**

Below are the verbatim generated responses from the two GenAI models:

4.1   Gemini

Absolutely, here's a concise Python code that demonstrates recursion by calculating the factorial of a number:

```python
def factorial(n):
    """
    This function calculates the factorial of a number using recursion.
    """
    if n == 0:
        return 1
    else:
        return n * factorial(n-1)

# Print the factorial of 5
print(factorial(5))
```

In this code, the *factorial* function takes a number n as input. It has two parts:

- Base Case: If n is equal to 0, the function returns 1. This is the stopping condition for the recursion. Factorial of 0 is defined as 1.
- Recursive Case: If n is not 0, the function returns n multiplied by the factorial of $n-1$. This is the recursive call where the function calls itself with a smaller value of 'n'.

When you call *factorial*(5), the function breaks down the problem into smaller subproblems:

- $factorial(5) = 5 * factorial(4)$
- $factorial(4) = 4 * factorial(3)$
- ...
- $factorial(1) = 1 * factorial(0)$ (base case reached)

The recursion continues until it reaches the base case (factorial of 0), then the results are returned back up the chain, calculating the factorial of 5.

This is a simple example of recursion, but it demonstrates the core concept of a function calling itself to solve a problem by breaking it down into smaller, similar subproblems.

4.2   ChatGPT

Sure, here's a simple example of a recursive function in Python that calculates the factorial of a number:

```python
def factorial(n):
    if n == 0:
        return 1
    else:
        return n * factorial(n - 1)
```

```python
# Example usage:
number = 5
print(f"The factorial of {number} is {factorial(number)}")
```

In this code:

- The factorial function is defined to calculate the factorial of a number n.
- The base case (if $n == 0$:) returns 1, which is necessary to stop the recursion.
- The recursive case (else:) computes $n * factorial(n - 1)$ which calls the *factorial* function again with a smaller argument until it reaches the base case.

When you run this code with *number* = 5, it will output:

```
The factorial of 5 is 120
```

This demonstrates how recursion works by breaking down a problem into smaller, simpler instances of the same problem until a base case is reached.

4.3   Evaluation of Gemini and ChatGPT

Objective Metrics. Both Gemini and ChatGPT seems to be accurate in terms of factual correctness and semantic correctness. Both models generate relevant Python codes as instructed that demonstrate recursion, and both generated codes compiled and ran with the expected outputs and no errors. In terms of speed, from sending the prompt to the response fully finishing, ChatGPT was a lot faster. Gemini took 8.38 s, and ChatGPT only needed 3.69 s. For the format, both models generated both source codes and textual paragraphs explaining the source codes in detail, which do not fully align with what the prompt asked for, since the prompt only asked for "Python code" and not explanation in text format.

Subjective Metrics. Both Gemini and ChatGPT delivered a cohesive and logical response with professional tone that one would normally see in academia. Both responses are clear with no ambiguity. However, both responses do not satisfy the "verbosity" criteria fully. The prompt asked for "short, concise, functioning Python code". Although the code portions are short, both GenAI models generated very detailed and verbose explanation about how the recursive function work, including the base case and the recursive case, and also examples of what the output would look like once executed.

4.4   Survey Questionnaire

To better demonstrate our framework's performance on GenAI prompt-response effectiveness on prompts related to computer science (again for simplicity purpose), we have come up with more potential prompts that can be used for evaluation via our framework.

- Please generate a Python code snippet that demonstrates how a binary search algorithm works on a sorted list.
- Write a concise C++ program to implement the quicksort algorithm, ensuring it is clear and efficient.
- Generate a Java function that utilizes a stack data structure to evaluate a postfix expression (Reverse Polish Notation).
- Create a JavaScript code example that illustrates the use of a breadth-first search (BFS) algorithm in a graph.
- Provide a short C# script that demonstrates how to implement a linked list, including insertion and deletion of nodes.
- Write a Java function to find the shortest path in a weighted graph using Dijkstra's algorithm.
- Generate a C program that implements the merge sort algorithm and includes a brief explanation of its time complexity.
- Create a Ruby code snippet that demonstrates how to use a hash table to store and retrieve data efficiently.
- Write a MATLAB function that performs matrix multiplication and explain its computational complexity.
- Provide a Swift example of how to use dynamic programming to solve the knapsack problem.
- Generate a concise Kotlin program that demonstrates how to implement a trie (prefix tree) for storing strings.
- Write a Rust function that uses recursion to solve the Tower of Hanoi problem and include an explanation of the solution.
- Create a JavaScript script that demonstrates the use of depth-first search (DFS) in finding connected components in an undirected graph.
- Provide a short Go code example that shows how to implement a binary search tree (BST) with insert and search operations.
- Write a Haskell function that implements the Floyd-Warshall algorithm for finding shortest paths in a weighted graph.

Note that conducting these surveys and interviews require human subject participation, and hence approval from Institutional Review Boards (IRBs) at universities and institutions.

5   Conclusion and Future Works

In this paper, we presented holistic and comprehensive framework for evaluating the effectiveness of prompts and responses. It integrates both objective metrics and subjective metrics to ensure a balanced assessment for both the technical performance of the AI and the nuanced human interaction experience with the users. Through the application of our evaluation framework (as demonstrated in our sample evaluation using Gemini and ChatGPT), GenAI researchers and developers can address the current limitations in GenAI and refine the AI generated content by better understanding the strengths and weaknesses of their respective GenAI model according to our evaluation framework. We can incorporated multiple feedback mechanisms that enable GenAI researchers and developers to use the evaluated metrics directly to improve GenAI through RLHF.

In order to fully evaluate the framework, future research work can be done to evaluate the GenAI responses to carefully designed prompts in all aspects and not just computer science, using our proposed framework. For our purpose, we suggest using two different GenAI models on the same sets of standardized prompts to ensure that the GenAI responses are comparable. Relevant objective and subjective metrics collected and measured using both interviews and surveys on different users as described in Sect. 3 to fully evaluate and improve the prompt-response effectiveness. Ultimately, our goal is to contribute to the development of GenAI systems that are not only technologically advanced but also trusted and valued by their users.

References

1. Ali, I., Ahmad, S., Usama, M., Muhammad, M.: ChatGPT: a comprehensive review on background, applications, key challenges, bias, ethical implications, limitations, and future directions. J. Innov. Digit. Ecosyst. **3**, 121–154 (2023)
2. Amoozadeh, M., et al.: Trust in generative AI among students: an exploratory study. arXiv preprint arXiv:2310.04631v2 (2023)
3. Argyle, L.P., Busby, E.C., Fulda, N., Gubler, J.R., Rytting, C., Wingate, D.: Out of one, many: using language models to simulate human samples. arXiv preprint arXiv:2209.06899 (2022)
4. Bandi, A., Adapa, P.V.S.R., Kuchi, Y.E.V.P.K.: The power of generative AI: a review of requirements, models, input–output formats, evaluation metrics, and challenges. Future Internet **15**(8), 260 (2023)
5. Bandi, A., Kagitha, H.: A case study on the generative AI project life cycle using large language models. In: Bandi, A., Hossain, M., Jin, Y. (eds.) Proceedings of 39th International Conference on Computers and Their Applications, EPiC Series in Computing, vol. 98, pp. 189–199. EasyChair (2024)
6. Bender, E.M., Gebru, T., McMillan-Major, A., Shmitchell, S.: On the dangers of stochastic parrots: Can language models be too big? In: Proceedings of the 2021 ACM Conference on Fairness, Accountability, and Transparency (FAccT 2021), pp. 610–623, New York, NY, USA (2021). Association for Computing Machinery
7. Bharti, S.K., et al.: Text-based emotion recognition using deep learning approach. Comput. Intell. Neurosci. **2022**, 2645381 (2022)

8. Centers for disease control and prevention. CDC museum Covid-19 timeline 2023. Accessed 11 June 2024

9. Ekin. S.: Prompt engineering for ChatGPT: a quick guide to techniques, tips, and best practices. TechRxiv, Version 2 (2023)

10. Hugging Face: Illustrating reinforcement learning from human feedback (RLHF). Hugging Face Blog (2023)

11. Giray, L.: Prompt engineering with ChatGPT: a guide for academic writers. Ann. Biomed. Eng. **51**(12), 2629–2633 (2023)

12. Google. Google Gemini API documentation (2023). https://ai.google.dev/gemini-api/docs. Accessed 12 June 2024

13. Heston, T.F., Khun, C.: Prompt engineering in medical education. Med. Educ. J. **2**(3), 19–25 (2023)

14. Huyen. C.: RLHF: reinforcement learning from human feedback (2023). Accessed 11 June 2024

15. Jeong, D., Aggarwal, S., Robinson, J., Kumar, N., Spearot, A., Park, D.S.: Exhaustive or exhausting? Evidence on respondent fatigue in long surveys. J. Dev. Econ. **161**, 102992 (2022)

16. Kim, Y., Lee, J., Kim, S., Park, J., Kim, J.: Understanding users' dissatisfaction with ChatGPT responses: types, resolving tactics, and the effect of knowledge level. In: Proceedings of the ACM on Human-Computer Interaction, vol. 7, no. CSCW2 (2023)

17. Kirk, R., et al.: Understanding the effects of RLHF on LLM generalisation and diversity. arXiv preprint arXiv:2310.06452 (2023)

18. Meskó, B.: Prompt engineering as an important emerging skill for medical professionals: tutorial. J. Med. Internet Res. **25**(1), e50638 (2023)

19. Mungoli, N.: Exploring the synergy of prompt engineering and reinforcement learning for enhanced control and responsiveness in chat gpt. J. Electr. Electron. Eng. **2**(3), 201–205 (2023)

20. Naseem, U., Bandi, A., Raza, S., Rashid, J., Chakravarthi, B.R.: Incorporating medical knowledge to transformer-based language models for medical dialogue generation. In: Proceedings of the 21st Workshop on Biomedical Language Processing, pp. 110–115 (2022)

21. Neumeister, L.: Lawyers submitted bogus case law created by ChatGPT. a judge fined them $5,000. AP News (2023)

22. University of Michigan: Prompt literacy (2024). https://genai.umich.edu/resources/prompt-literacy. Accessed 12 June 2024

23. OpenAI. Openai platform: Text generation guide (2023). https://platform.openai.com/docs/guides/text-generation. Accessed 12 June 2024

24. Ouyang, L., et al.: Training language models to follow instructions with human feedback. arXiv preprint arXiv:2203.02155 (2022)

25. Stack Overflow. std::vector versus std::array in c++ (2023). Accessed 11 June 2024

26. Patel, H., Shah, K.A., Mondal, S.: Do large language models generate similar codes from mutated prompts? A case study of Gemini Pro. In: Proceedings of the ACM International Conference on the Foundations of Software Engineering (FSE) (2024)

27. Python software foundation. Sorting HOWTO (2023). Accessed 11 June 2024

28. Jack, W.R., et al.: Scaling language models: methods, analysis & insights from training gopher. arXiv preprint arXiv:2112.11446 (2021)

29. Shan, M., Zhang, W., Li, T., Wang, H., Sun, L.: Artificial hallucinations in ChatGPT: implications in scientific writing. Cureus **15**(2), e35021 (2023)

30. Short, C.E., Short, J.C.: The artificially intelligent entrepreneur: Chatgpt, prompt engineering, and entrepreneurial rhetoric creation. J. Bus. Ventur. Insights **19**, e00388 (2023)
31. Wang, J., et al.: The good, the bad, and why: unveiling emotions in generative AI. arXiv preprint arXiv:2312.11111 (2023)
32. White, J., et al.: A prompt pattern catalog to enhance prompt engineering with ChatGPT. Artif. Intell. Rev. **5**(2), 101–110 (2023)
33. Fangjun, Yu., Quartey, L., Schilder, F.: Exploring the effectiveness of prompt engineering for legal reasoning tasks. Find. Assoc. Comput. Linguist. ACL **2023**, 13582–13596 (2023)

Blockchain and Security

Securing RC Based P2P Networks: A Blockchain-Based Access Control Framework Utilizing Ethereum Smart Contracts for IoT and Web 3.0

Saurav Ghosh[1], Reshmi Mitra[1(✉)], Indranil Roy[1], and Bidyut Gupta[2]

[1] Southeast Missouri State University, Cape Girardeau, USA
{sghosh3s,rmitra,iroy}@semo.edu
[2] Southern Illinois University, Carbondale, USA
bidyut@cs.siu.edu

Abstract. Ensuring security for highly dynamic peer-to-peer (P2P) networks has always been a challenge, especially for services like online transactions and smart devices. These networks experience high churn rates, making it difficult to maintain appropriate access control. Traditional systems, particularly Role-Based Access Control (RBAC), often fail to meet the needs of a P2P environment. This paper presents a blockchain-based access control framework that uses Ethereum smart contracts to address these challenges. Our framework aims to close the gaps in existing access control systems by providing flexible, transparent, and decentralized security solutions. The proposed framework includes access control contracts (ACC) that manage access based on static and dynamic policies, a Judge Contract (JC) to handle misbehavior, and a Register Contract (RC) to record and manage the interactions between ACCs and JC. The security model combines impact and severity-based threat assessments using the CIA (Confidentiality, Integrity, Availability) and STRIDE principles, ensuring responses are tailored to different threat levels. This system not only stabilizes the fundamental issues of peer membership but also offers a scalable solution, particularly valuable in areas such as the Internet of Things (IoT) and Web 3.0 technologies.

1 Introduction

Studying security issues in peer-to-peer (P2P) networks is crucial due to their inherent vulnerabilities arising from decentralized control, which makes them prime targets for various attacks such as Sybil, man-in-the-middle, and denial-of-service (DoS) attacks. In P2P networks, each peer acts as both a client and server, creating numerous points of potential failure and compromise. Traditional role-based access control (RBAC) models are inadequate because they are designed for centralized systems where roles and permissions are clearly defined and controlled by a central authority. The lack of centralization in P2P networks makes it difficult to enforce consistent access controls and manage identities securely.

One significant issue with RBAC in P2P networks is Role Explosion, where managing multiple roles for all possible entities becomes unmanageable. The

G. Hu et al. (Eds.): CAINE 2024, CCIS 2242, pp. 73–85, 2025.
https://doi.org/10.1007/978-3-031-76273-4_6

static nature of roles in RBAC conflicts with the dynamic characteristics of P2P networks, where nodes frequently join and leave. This constant change leads to an explosion of roles, complicating the maintenance and updating of role definitions, and creating potential security gaps. Addressing security in P2P networks requires developing more robust and adaptable security mechanisms capable of handling their dynamic and decentralized nature, managing identities and access controls in real time, and overcoming the limitations of traditional RBAC models.

In this paper, we introduce a blockchain-based access control architecture that tackles these challenges by harnessing the properties of blockchain: constant verification, immutability, transparency, and decentralization. Blockchain's ongoing verification processes ensure that each transaction and access request is validated by multiple nodes, creating a resilient security framework suitable for dynamic and unsecured environments such as P2P networks [5,12]. Our proposed architecture consists of three main components:

- **Register Contract (RC):** This maintains a comprehensive registry of all access control methods and their corresponding smart contracts, making it easier to register, update, and delete methods. The concept of an RC was initially proposed by Zhang et al. [16]. Our work expands on it by incorporating dynamic policy management and real-time updates to better handle the fluid nature of P2P networks.
- **Judge Contract (JC):** This contract evaluates reported misbehaviors, determines penalties, and ensures appropriate actions are taken against violators, guaranteeing dynamic and fair adjudication of security breaches.
- **Access Control Contracts (ACCs):** These contracts implement specific access control policies for subject-object pairs. They perform static validation based on predefined policies and dynamic validation by monitoring subject behavior.

Our work stands out for its comprehensive approach to integrating blockchain technology with access control mechanisms. Unlike previous studies that mainly used blockchain for storing access policies or managing data records [5,12], our framework leverages the computational capabilities of smart contracts to enforce security policies dynamically. This dual-layer approach significantly enhances security, scalability, and adaptability in decentralized networks. The main contributions of this paper are:

- **Blockchain-Based Access Control Framework:** We developed a new access control framework using Ethereum smart contracts to improve security in decentralized networks.
- **Policy Management:** We introduced mechanisms for effective policy management within the framework, capable of adapting in real-time to network conditions and security threats.
- **Scalable Security Solutions for Decentralized Networks:** We introduced a blockchain-based platform design for IoT and Web 3.0 solutions, allowing secure decentralized system growth.

– **Empirical Validation and Practical Application:** Demonstrated usefulness through practical application, showcasing potential for large-scale adoption.

This paper has six main sections. In Sect. 2, the summary of the access control and blockchain applications in P2P networks is presented. Section 3 describes the RC-based P2P architecture. Section 4 describes the design, whereas Sect. 5 describes the implementation of the blockchain-based access control framework, including the integration of the Ethereum smart contracts. Section 6 discusses the results and evaluation of the framework's impact on network security through several deployments. Finally, Sect. 7 concludes with the accomplishments of this research.

2   Related Work

Security in IoT and P2P Networks: The need for robust security frameworks in IoT environments is critical. Novo [10] proposed scalable blockchain architecture for IoT access management, demonstrating how blockchain can enhance security and operational efficiency. Yu, Chen, and Wang [15] highlighted additional security challenges, such as latency and node heterogeneity in distributed networks. Mbarek, Ge, and Pitner [8] extended these concepts to smart home and smart grid systems, emphasizing the improved data integrity and resilience against cyber-attacks within peer-to-peer networks.

Blockchain-Based Access Control Mechanisms: Sun et al. [13] reviewed the transformative role of blockchain in IoT access control, focusing on enhanced transparency through smart contracts. Ding et al. [3] proposed an attribute-based access control scheme that minimizes overhead while enhancing security and practicality. Similarly, Wang et al. [14] developed a dynamic access control system that adapts to user behavior and device status in real-time, showcasing blockchain's flexibility in complex network environments. Liu et al. [7] demonstrated the impact of integrating blockchain into IoT networks to enhance access control mechanisms, thereby reducing risks from unauthorized access and breaches. Practical examples provided by Sultana et al. [11] and Badhe and Arjunwadkar [1] illustrated the scalability and adaptability of blockchain-based smart contracts for secure data sharing and distribution in contemporary IoT frameworks.

Collaborating access control systems with blockchain models in P2P networks offers a robust solution for IoT security. Zhang et al. [17] provided a critical evaluation of various blockchain-based models, identifying key areas for scalability and security improvements. Han et al. [4] and Li et al. [6] further elaborated on enhancing access control through dual-layer blockchain systems, promoting reliability and flexibility.

Our Work Builds on These Foundations by Integrating Dynamic Policy Management and Real-Time Updates. This means our approach can easily adapt to the ever-changing nature of P2P networks. This advancement is a game-changer.

It shows just how practical and applicable our proposed framework is on a large scale, making a significant contribution to the field of decentralized network security. Our framework goes above and beyond by tackling the limitations of existing models and boosting the overall security and operational integrity of decentralized environments.

3   Preliminaries

We have taken into consideration some of the first results of an RC-based low-diameter two-level hierarchical structured P2P network [9], which is a two-level overlay architecture, and at each level structured networks of peers exist. It is explained in detail below.

1. At level-1, we have a ring network consisting of the peers H_i $(0 \leq i \leq n-1)$. The number of peers on the ring is n, the number of distinct resource types. This ring network is used for efficient data lookup, so we name it a transit ring network.
2. At level 2, there are n numbers of completely connected networks (groups) of peers. Each such group, say G_i is formed by the peers of the subset P^{Ri}, $(0 \leq i \leq n-1)$, such that all peers $(\in P^{Ri})$ are directly connected (logically) to each other, resulting in the network diameter of 1. Each G_i is connected to the transit ring network via its group-head H_i.
3. Each peer in the network maintains an Information Resource Table (IRT) that consists of n number of tuples.

4   Design of Access Control Framework

Blockchain-based access control distributes the validation and enforcement of policies across multiple nodes in the network. This means that access requests and transactions are continuously verified by multiple participants, which ensures security and trust among network members and eliminates the risk of a single point of failure. The proposed framework consists of multiple access control contracts (ACCs), a Judge Contract (JC), and a Register Contract (RC). Each network entity, such as group heads and members, is assigned access privileges based on roles and responsibilities. These privileges are managed at a granular level by various ACCs, while the RC [2,5,12] maintains a system-level view of all access control policies. The JC receives regular monitoring updates from ACCs about various STRIDE events and issues penalties accordingly. Our design philosophy focuses on *flexibility, scalability, and robustness*. The architecture dynamically manages access permissions and enforces security policies in real time, accommodating high churn rates and the structural openness of P2P networks. Using Ethereum smart contracts, our framework can implement complex access control policies and respond effectively to security threats (Figs. 1, 2).

4.1   Access Control Contracts (ACCs)

ACCs are the core components of the access control framework, implementing specific policies for subject-object pairs. These contracts manage various tasks within the P2P network such as viewing, editing, and managing the Global Resource Table (GRT), accessing local resources, and modifying group configurations.

A dual validation mechanism is implemented to enhance security further. This mechanism includes:

– **Static Access Right Validation:** This layer checks predefined policies against access requests to ensure that all requests comply with granted permissions and are managed securely.

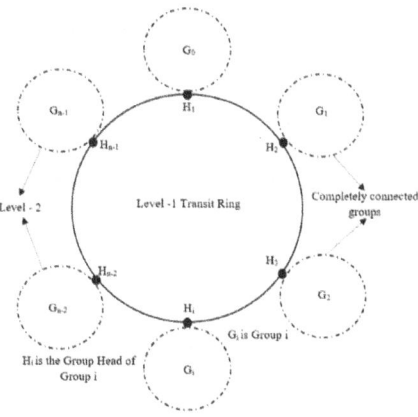

Fig. 1. A two-level RC based structured P2P architecture with n distinct resource types

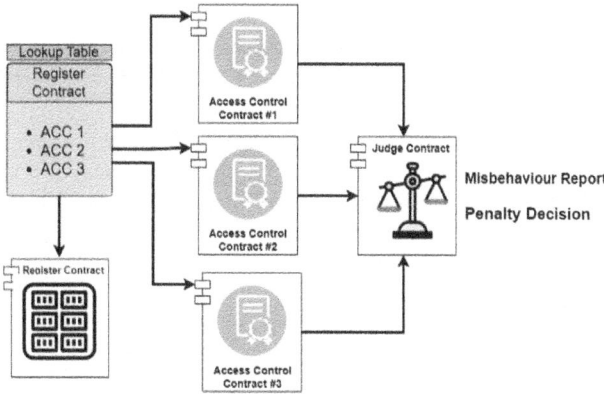

Fig. 2. Architecture of the Access Control System

- **Dynamic Access Right Validation:** It evaluates the subject's behavior in real-time, adjusting access rights dynamically in response to emerging threats or abuses.

Additionally, each ACC maintains a misbehavior list detailing actions considered inappropriate, the timing of such actions, and the applied penalties. This record allows for dynamic adjustments to access permissions and maintains the discipline within the P2P system.

1. **Storage**: All ACCs and their policies are stored on the Ethereum blockchain, ensuring immutability and transparency.
2. **Decision Making**: Call acceptance and validation are decided by the collective consensus of the network nodes executing the smart contract code.
3. **Failure vs Success**:
   - **Failure**: If the subject's request does not comply with predefined policies or fails adaptive validation due to suspicious behavior, the access request is denied, and the transaction is recorded on the blockchain.
   - **Success**: If the subject's request meets all validation criteria, access is granted, and the transaction is recorded on the blockchain.
4. **Sample Tasks**: Examples of tasks managed by ACCs include:
   - **View GRT**: Allowing a subject to view the Global Resource Table.
   - **Edit GRT**: Granting permissions to modify entries in the Global Resource Table.
   - **Access Local Resources**: Enabling access to local resources based on the subject's role.
   - **Modify Group Configurations**: Allowing changes to group member roles and resource allocations.

This table shows a representation of the permissions that different peers have.

Table 1. Static Access permissions of different Subject

Roles / Access	Global Resource Table	Local Resource Table	Malicious Group Head	Malicious Member	Communication Access
Primary Group Head	full (view, edit, create, delete)	full (view, edit, create, delete)	full (view, edit, create, delete)	full (view, edit, create, delete)	other group head, own members
Secondary Group Head	view	view	view, edit	deny	own group head, own members
Regular Members	deny	view	deny	deny	own group head

Table 2 presents a range of security policies designed to protect P2P systems by dynamically responding to different types of security violations.

Table 2. Security event responses based on severity and impact

Security Event	Severity	CIA Impact	STRIDE Impact	Response
Too many Access Attempts	High	Integrity	Elevation of Privilege	24-hour ban
Data Tampering	High	Integrity	Tampering	Cancel access permanently
Unauthorized Access	Medium	Confidentiality	Information Disclosure	Temporary suspension
Disruption of Service	High	Availability	Denial of Service	30 d suspension
Misrepresentation of Identity	Low	Accountability	Spoofing	Warning, penalties

4.2   Judge Contract (JC)

The JC judges the misbehavior of the subject and determines the corresponding penalty when receiving a potential misbehavior report from an ACC. After determining the penalty, the JC returns the decision to the ACC for further operation.

- **Object**: The peer who experienced the misbehavior. For example, a Primary Group Head (PGH) who noticed unauthorized access attempts to the Global Resource Table (GRT).
- **Misbehavior**: Detailed description of the misbehavior. For instance, multiple failed login attempts by a Secondary Group Head (SGH) trying to access restricted resources.
- **Time**: The specific time when the misbehavior occurred. An example would be a timestamp indicating when a Regular Member repeatedly tried to access Global Resource Table's Data.
- **Penalty**: The penalty imposed based on the severity of the misbehavior. For example, a temporary suspension of access rights for 24 h for the Secondary Group Head due to multiple failed login attempts.

4.3   Register Contract (RC)

The Register Contract provides a centralized yet decentralized repository for all access control methods, ensuring that each method can be easily registered, updated, or deleted. It offers a system-level view of the entire network access control policies, which also provides dynamic reference data to different roles within the network, facilitating secure and efficient interactions.

The RC will also maintain a lookup table, which should look like this:

- MethodName: the name of the method;
- Subject: the subject of the corresponding subject-object pair of the method;
- Object: the object of the corresponding subject-object pair of the method;
- ScName: the name of the corresponding smart contract implementing this method;
- Creator: the peer who created and deployed the contract;
- ScAddress: the address of the smart contract;

The RC provides the following main ABIs to manage these methods.

- `methodRegister()`: Verify a new method and register the information into the lookup table.
- `methodUpdate()`: Verification of an existing method that needs to be updated and updates the information, especially the fields of ScAddress and ABI.
- `methodDelete()`: Verifies the MethodName of a method and deletes the method from the lookup table.
- `getContract()`: Verifies the MethodName of a method and returns the address and ABIs of the contract (i.e., the ACCs and JC) of the method.

The RC is also engineered to dynamically provide reference data to different roles within the network based on their needs. For instance:

- When a Primary Group Head acts as the subject and requests access to a resource or interaction, the RC supplies the reference data, such as the "Primary Group Head Role ACC," to facilitate the access or communication.
- Similarly, if a Secondary Group Head or Regular Member assumes the role of the subject, intending to connect with a resource or another entity, the RC provides them with their respective ACC references, like "Secondary Group Head Role ACC" or "Regular Members Role ACC," accordingly.

This dynamic reference mechanism ensures that each peer, depending on their role and the nature of the access request, is equipped with the appropriate smart contract references. It facilitates secure interaction within the network, ensuring that access controls are adhered to the predefined policies and permissions (Table 3).

In our blockchain-based system, the communication between primary group heads (PGHs) is efficiently managed through a series of interactions with smart contracts, specifically the Registry Contract and the Primary Group Head Role Access Control Contract (ACC). Detailed steps of the communication process are fully illustrated in the accompanying diagrams, which depict how access requests are handled, validated, and logged to maintain network security and integrity (Fig. 3).

Table 3. Dynamic reference data supplied by the RC

Subject Role	Reference Data Supplied by RC
Primary Group Head	Primary Group Head Role ACC
Secondary Group Head	Secondary Group Head Role ACC
Regular Member	Regular Member Role ACC

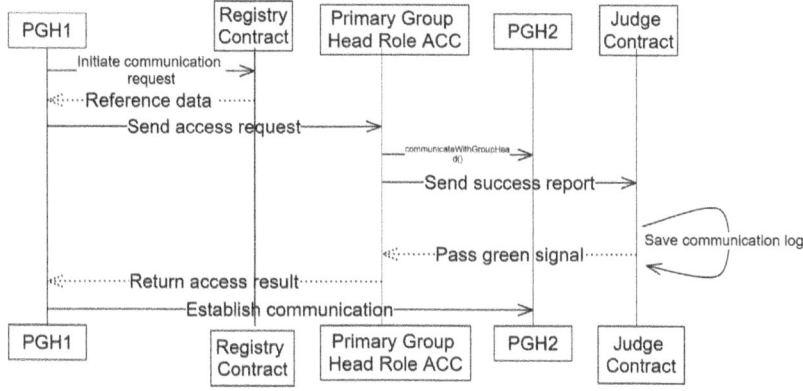

Fig. 3. Blockchain-based access control system architecture

The blockchain-based access control system is designed to tightly regulate any updates or deletions made to Access Control Contracts (ACCs), ensuring that any modifications meet strict security and integrity standards. Whenever there are updates to the access control methods, they go through a thorough review process to make sure they're compatible with predefined roles and the overall system architecture. This helps maintain the security of the network and prevents any unauthorized access.

Two key contracts in our system are the Register Contract (RC) and the Judge Contract (JC). These contracts have important roles in overseeing the application of policies and the execution of roles within the network. They make sure that there are secure interactions among the Primary Group Heads (PGHs) and that the access control mechanisms are tailored to the specific needs of the network, ultimately improving operational efficiency.

The novel aspect of our work lies in the integration of blockchain technology with a dynamic access control mechanism that uses dual-validation. This mechanism is specifically designed for peer-to-peer (P2P) networks. By distributing the validation process across multiple nodes, the framework basically eliminate any single points of failure and continuously verify access requests and transactions. The system also adapts in real-time by adjusting permissions based on user behavior, allowing for immediate responses to any potential threats. By storing the ACCs and policies on the Ethereum blockchain, the system ensure that they cannot be changed or tampered with.

The dynamic reference data mechanism provided by the Register Contract adds role-specific access controls, which improves communication and overall efficiency. The architecture is built to handle high churn rates and the structural openness of P2P networks, making it scalable and flexible for networks of different sizes and structures.

In sum, the access control system, which is based on blockchain technology, along with the strict operational guidelines and the important roles played by the RC and JC, ensures a secure and efficient network. This approach not only improves communication and operational efficiency, but also maintains a high standard of security and integrity, which is crucial for the reliable operation of blockchain-based systems.

4.4 Algorithm

The access control logic defines roles and permissions for different group members, implementing functions for resource access and management. Below is the algorithm used to control access based on the role and resource table. The lines 2,7,23,19 correspond to the policies outlined in Table 1.

Algorithm 1. Access Control Logic

 1: **function** ACCESSCONTROL(*resourceTable*)
 2:     **if** *resourceTable* == Global Resource Table, Local Resource Table **then**
 3:         **return** Full access: view, edit, create, delete
 4:     **end if**
 5:
 6:     **function** NORMALPRIMARYGH(*resourceTable*)
 7:         **if** *resourceTable* == Global Resource Table, Local Resource Table **then**
 8:             **return** view
 9:         **end if**
10:     **end function**
11:
12:     **function** NORMALSECONDARYGH(*resourceTable*)
13:         **if** *resourceTable* == Global Resource Table, Local Resource Table **then**
14:             **return** view
15:         **end if**
16:     **end function**
17:
18:     **function** NORMALREGULARMEMBER(*resourceTable*)
19:         **if** *resourceTable* == Local Resource Table **then**
20:             **return** view
21:         **else**
22:             **return** deny
23:         **end if**
24:     **end function**
25: **end function**

The main function is `AccessControl`, which is called by the respective ACC based on the role and resource being accessed. For instance, when a Primary Group Head (PGH) attempts to access the resource table, the ACC invokes the `NormalPrimaryGH` function to determine the appropriate access level. This call is made dynamically by the smart contract handling the access request.

5   Implementation

Hardware Used: We used HP Probook G4 440, which is equipped with an Intel(R) Core(TM) i7-7500U CPU, running at a frequency of 2.70GHz with 2 cores and 4 logical processors. Operating on Microsoft Windows 11 Pro (64 bit), the device has a memory capacity of 15.9 GB. For storage, it offers a 931.51 GB HDD and a 223.57 GB SSD.

The various contracts implemented within our blockchain-based access control system, each tailored to specific roles and their corresponding access levels as depicted in Table 1 of Subsect. 4.1.

BaseAccessControl Contract: This foundational contract holds basic resource management and internal communication. Serving as a template, it allows for modular extensions specific to each role, laying out a framework where general permissions are specified further.

PrimaryGroupHeadRoleACC Contract: Extending from `Base AccessControl`, this contract provides Primary Group Heads with comprehensive resource management rights. It effectively translates the conceptual "Full access" into practical abilities to view, edit, create, and delete resources both locally and across the network.

SecondaryGroupHeadRoleACC Contract: Tailored for Secondary Group Heads, this contract confines its functionalities to viewing resources only, aligning with the prescribed limited access. It enforces strict adherence to the hierarchical access control by prohibiting any modifications to resources.

RegularMembersRoleACC Contract: This contract limits Regular Members to accessing local resources only, consistent with the stringent access controls outlined by the algorithm.

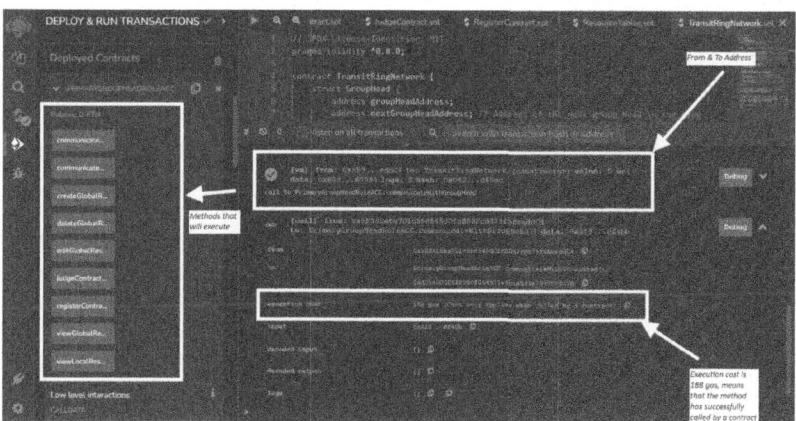

Fig. 4. Description of the PrimaryGroupHeadRoleACC contract and the associated transaction.

At the bottom, we have the input data, followed by the decoded inputs, outputs, and logs from the function, indicating that the execution was successful. As shown in Fig. 4, this process confirms the successful execution.

6   Result and Discussion

The blockchain-based access control framework for P2P networks applied in this case has resulted in advanced levels of network security and effectiveness as we can see in the comparison of system performance Before and after the Implementation below:

Category	Before Implementation	After Implementation
Response Times	Slow	Improved
Unauthorized Access	High rates	Reduced
Access Management Transparency	Lack of transparency	Enhanced

7   Conclusion

This paper presents a blockchain-based access control framework for high-security peer-to-peer (P2P) networks, specifically in IoT and Web 3.0 environments. Tested in an RC-based P2P network, it leverages Ethereum smart contracts for immutable, transparent, decentralized security management, reducing vulnerabilities. The novelty of the work lies in integrating blockchain with dynamic, dual-validation access control, enhancing security by distributing validation across nodes, verifying access requests, and eliminating single points of failure. Real-time adaptability ensures immediate threat response, while immutable Ethereum records build trust.

Future work includes developing additional Access Control Contracts (ACCs) to handle other functions and roles within the network, further enhancing the granularity and security of our access control system.

References

1. Badhe, G., Arjunwadkar, D.M.: Access control systems based on blockchain technology. Int. J. Eng. Technol. Manag. Sci. (2023)
2. Dhif, Y., Ayed, H.K.B., Zaatouri, K.: A blockchain based access control for IoT. In: 2019 15th International Wireless Communications & Mobile Computing Conference (IWCMC). IEEE (2019)
3. Ding, N., Cao, J., Li, C., Fan, K., Li, H.: A novel attribute-based access control scheme using blockchain for IoT. IEEE Access (2019)

4. Han, D., Zhu, Y., Li, D., Liang, W., Souri, A., Li, K.-C.: A blockchain-based auditable access control system for private data in service-centric IoT environments. IEEE Trans. Ind. Inf. (2022)
5. Islam, M.A., Madria, S.: A permissioned blockchain based access control system for IoT. In: 2019 IEEE International Conference on Blockchain. IEEE (2019)
6. Li, Z., Hao, J., Liu, J., Wang, H., Xian, M.: An IoT-applicable access control model under double-layer blockchain. Express briefs, IEEE Transactions on Circuits and Systems II (2021)
7. Liu, H., Han, D., Li, D.: Fabric-IoT: a blockchain-based access control system in IoT. IEEE Access (2020)
8. Mbarek, S., Ge, M., Pitner, T.: Blockchain-based access control for IoT in smart home systems. In: Proceedings of the 31st International Conference on Database and Expert Systems Applications (DEXA 2020). IEEE (2020)
9. Neupane, A., Mitra, R., Roy, I., Gupta, B., Debnath, N.: Efficient and secured data lookup protocol using public-key and digital signature authentication in RC-based hierarchical structured P2P network. Int. J. Comput. Their Appl. **30**(2), 140 (2023)
10. Novo, O.: Blockchain meets IoT: an architecture for scalable access management in IoT. IEEE Internet Things J. **5**(2), 1184–1195 (2018)
11. Sultana, T., Almogren, A.S., Akbar, M., Zuair, M., Ullah, I., Javaid, N.: Data sharing system integrating access control mechanism using blockchain-based smart contracts for IoT devices. Appl. Sci. (2020)
12. Sun, S., Chen, S., Du, R., Li, W., Qi, D.: Blockchain based fine-grained and scalable access control for IoT security and privacy. In: 2019 IEEE Fourth International Conference on Data Science in Cyberspace (DSC), pp. 598–603. IEEE (2019)
13. Sun, Y., Song, H., Tu, Z., Qin, Y.: Blockchain for IoT access control: recent trends and future research directions. Sensors (Basel, Switzerland) (2022)
14. Wang, Y., Xu, N., Zhang, H., Sun, W., Benslimane, A.: Dynamic access control and trust management for blockchain-empowered IoT. IEEE Internet Things J. (2022)
15. Yu, H., Chen, T., Wang, J.: A blockchain-based access control mechanism for IoT. In: Proceedings of the 2022 6th International Conference on Electronic Information Technology and Computer Engineering. IEEE (2022)
16. Zhang, X.X. et al.: Smart contracts for access control: can blockchain and cryptography support a more secure framework? J. Crypt. **12**, 345–367 (2018.) https:// allquantor.at/blockchainbib/pdf/zhang2018smart.pdf
17. Zhang, Y., Memariani, A., Bidikar, N.: A review on blockchain-based access control models in IoT applications. In: 2020 IEEE 16th International Conference on Control & Automation (ICCA). IEEE (2020)

Integrating Blockchain-Based Security and Privacy with QML in Edge Computing for 6G Networks

Jongho Seol[1] (ID), Jongyeop Kim[2]([✉]) (ID), and Abhilash Kancharla[3] (ID)

[1] Middle Georgia State University, Macon, GA 31206, USA
jongho.seol@mga.edu
[2] Georgia Southern University, Statesboro, GA 30460, USA
jongyeopkim@georgiasouthern.edu
[3] The University of Tampa, Tampa, FL 33606, USA
akancharla@ut.edu

Abstract. This paper presents a robust theoretical framework to integrate blockchain-based security and privacy mechanisms with quantum machine learning (QML) in the edge computing domain within 6G networks. The burgeoning deployment of edge devices necessitates secure, privacy-preserving, and trustworthy infrastructures to support collaborative QML tasks while upholding data confidentiality at the network periphery. Leveraging blockchain technology's decentralized and immutable ledger capabilities, this framework manages access control, ensures data provenance, and verifies integrity in edge computing environments. Furthermore, integrating quantum-resistant cryptographic primitives is explored to fortify defenses against potential threats from quantum adversaries. In addition to these considerations, the paper incorporates the theory of quantum probability within the framework, particularly in the context of the central limit theorem, to account for the probabilistic nature of quantum systems and its implications on statistical inference in QML tasks. Detailed latency analysis reveals that blockchain processing time increases with transaction complexity, quantum processing time grows more slowly, and 6G transmission time remains constant due to high bandwidth capabilities. Incorporating machine learning components such as data preprocessing and model inference times provides a comprehensive understanding of edge computing performance. Combining blockchain-based security and privacy measures with QML techniques like federated learning and differential privacy, the envisioned framework strives to establish a secure and trusted ecosystem for collaborative QML tasks at the network edge. This theoretical endeavor, enriched by quantum probability theory and detailed latency analysis, lays a solid groundwork for further research and development in this burgeoning interdisciplinary domain, promising advancements in edge computing applications' efficiency, reliability, and security within future wireless communication infrastructures.

Keywords: Blockchain · Quantum Machine Learning (QML) · Edge Computing · 6G Networks · Security

G. Hu et al. (Eds.): CAINE 2024, CCIS 2242, pp. 86–101, 2025.
https://doi.org/10.1007/978-3-031-76273-4_7

1 Introduction

The advent of 6G networks [1] promises a paradigm shift in wireless communication, ushering in an era of unprecedented connectivity, ultra-low latency, and massive device connectivity. Edge computing [17] is at the forefront of this evolution, a distributed computing paradigm that brings computation and data storage closer to end-users and devices. In conjunction with the capabilities of 6G networks, edge computing can revolutionize various sectors, including healthcare, transportation, manufacturing, and smart cities, by enabling real-time data processing, low-latency applications, and personalized services [18]. However, the widespread adoption of edge computing within 6G networks introduces new security, privacy, and trust challenges. As edge devices become increasingly interconnected and heterogeneous, there is a growing need for robust mechanisms to safeguard sensitive data, ensure data privacy, and establish trust in the integrity of computational processes. In response to these challenges, this paper proposes a novel theoretical framework that integrates blockchain-based security and privacy mechanisms with quantum machine learning (QML) [4, 5] techniques tailored for edge computing environments within 6G networks [20]. By leveraging the decentralized and immutable nature of blockchain technology, combined with the computational power and privacy-enhancing features of quantum machine learning, our framework aims to address the inherent security and privacy concerns while unlocking the potential of edge computing in the 6G era. Through this paper, we aim to elucidate the theoretical underpinnings of our proposed framework, explore its practical implications through case studies and experiments, and contribute to the ongoing discourse on the future of edge computing within the context of 6G networks.

Edge computing has emerged as a critical enabler of 6G networks, facilitating real-time data processing and analysis at the network edge. By offloading computation and storage tasks from centralized data centers to edge devices, edge computing reduces latency, improves response times, and enhances user experience [19]. However, the distributed nature of edge computing introduces new challenges in ensuring security and privacy, particularly in environments with resource-constrained devices and heterogeneous networks.

In 6G networks, edge computing brings data processing and computational power closer to the end-users by utilizing edge nodes strategically positioned near the data sources. These edge nodes, often embedded in base stations, local servers, or even user devices, handle data locally, which significantly reduces the latency and bandwidth usage compared to traditional cloud computing. By processing data at the network's edge, closer to where it is generated, edge computing allows for real-time analytics and decision-making. This local processing is crucial for applications requiring ultra-low latency and high reliability, such as autonomous vehicles, augmented reality (AR), and real-time video analytics. The decentralized nature of edge computing also enhances security and privacy, as data does not need to travel to distant central servers, thereby reducing the risk of interception and exposure. This approach improves performance and user experience and optimizes network resources, making 6G networks more efficient and responsive to the growing demands of modern applications.

Quantum machine learning harnesses the computational power of quantum computers to tackle complex optimization and machine learning tasks. QML algorithms can revolutionize data analysis, pattern recognition, and decision-making in 6G networks. However, integrating QML with existing security and privacy mechanisms poses unique challenges due to the probabilistic nature of quantum systems and the need for quantum-resistant cryptography.

The primary objective of this paper is to develop a robust theoretical framework that integrates blockchain-based security and privacy mechanisms with quantum machine learning in the edge computing domain within 6G networks. This framework aims to support secure, privacy-preserving, and trustworthy collaborative QML tasks, addressing the data confidentiality challenges posed by the increasing deployment of edge devices.

This paper is structured to methodically explore the integration of blockchain security and privacy with quantum machine learning (QML) in edge computing for 6G networks [20]. Starting with an Introduction that outlines the necessity for advanced solutions in 6G, it moves to a Literature Review summarizing existing studies and identifying research gaps. The Research Problem section delineates the specific challenges addressed by this integration. The Model Architecture and Contribution section provides an in-depth look at the proposed system, supported by diagrams and specifications, and presents novel strategies to merge these technologies. Analysis of Simulation Results follows, showcasing the performance improvements through simulations. Finally, the Conclusion encapsulates the findings and suggests future research directions.

2 Literature Review

The evolution of wireless communication networks from 1G to 6G represents a journey marked by significant technological advancements and paradigm shifts. As we approach the era of 6G networks [1], characterized by ultra-high data rates, ultra-low latency, and massive device connectivity, the role of edge computing has become increasingly prominent. Edge computing, a distributed computing paradigm that brings computation and data storage closer to the end-users and devices, offers a solution to the growing demand for real-time data processing and analytics. In today's digital landscape, where data is generated at an unprecedented rate from various sources such as sensors, mobile devices, and IoT endpoints, traditional cloud-based approaches to data processing and analysis are often insufficient to meet the stringent latency requirements of emerging applications. Edge computing addresses this challenge by moving computation and data storage closer to the data source, thereby reducing the distance data needs to travel and minimizing latency. Recent literature has extensively explored the potential of edge computing in 6G networks, highlighting its ability to not only reduce latency and improve response times but also support emerging applications such as augmented reality (AR), virtual reality (VR), and Internet of Things (IoT) devices [3]. By offloading computation and data processing tasks from centralized cloud servers to edge devices closer to the data source, edge computing enables real-time data analysis and decision-making, paving the way for innovative applications and services requiring low-latency and high-throughput data processing capabilities.

Moreover, the distributed nature of edge computing architectures enhances scalability, reliability, and resilience, making them well-suited for dynamic and heterogeneous environments characteristic of 6G networks. As 6G networks continue to evolve and proliferate, the role of edge computing is poised to become even more critical, driving innovation and enabling new use cases across various industries and sectors. The illustration of the architecture of 6G [2] is shown in Fig. 1.

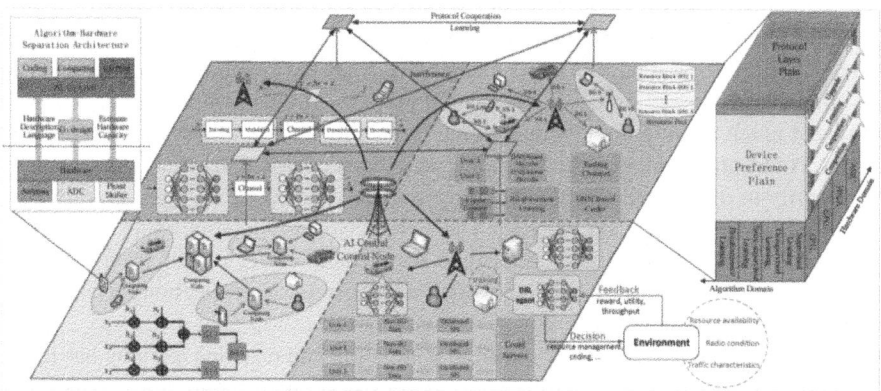

Fig. 1. The architecture of 6G [2].

The architecture of 6G, shown in Fig. 1, is poised to revolutionize telecommunications by leveraging advancements across various technologies [2]. It will employ terahertz communication, utilizing frequencies from 0.1 to 10 THz, to achieve ultra-high-speed data transmission, suitable primarily for short-range communications due to high attenuation. Artificial intelligence and machine learning will be deeply integrated into network management, automating operations, predicting issues, and enhancing security, with edge AI processing data closer to the source for real-time analytics and decision-making. Advanced network slicing will allow the creation of multiple virtual networks tailored to different applications, dynamically adjusting resources based on demand and service requirements. Intelligent and meta- surfaces will direct and reflect signals efficiently, improving coverage and capacity in challenging environments. Security will be bolstered through decentralized blockchain technologies and enhanced privacy measures, ensuring secure, transparent, and tamper-proof data exchanges. Key features of 6G will include ultra-low latency, targeting as low as 1 ms for critical applications such as autonomous driving and remote surgery; massive connectivity, supporting up to 10 million devices per square kilometer for IoT and smart cities; and high reliability, ensuring 99.999% uptime for mission-critical applications. Additionally, energy efficiency will be a priority, with innovations aimed at reducing the network's environmental impact and operational costs. Deployment will require dense networks with more base stations and small cells, advanced fiber optic and satellite backhaul solutions, and collaborative efforts from international standards bodies for global compatibility. Spectrum allocation will also be crucial, with regulatory bodies managing high-frequency spectrum to avoid interference and maximize utilization. Overall, the 6G architecture aims to create a more intelligent, flexible, and efficient network to meet the demands of future applications and services.

Various theoretical frameworks have been proposed to address security in machine learning implementations over networks, particularly in the context of 6G. Conventional approaches often rely on cryptographic techniques such as homomorphic encryption to protect data privacy and integrity during processing [21]. Homomorphic encryption allows computations on encrypted data without decryption, ensuring confidentiality. Federated learning is another framework where models are trained across decentralized devices holding local data samples, enhancing privacy by keeping raw data localized and only sharing model updates. Differential privacy ensures computation outputs do not reveal much about any individual input, protecting data points even with multiple queries. Additionally, blockchain technology has been explored to ensure data integrity and security in distributed machine learning systems, providing a tamper-proof ledger to verify data provenance and integrity. These frameworks—homomorphic encryption, federated learning, differential privacy, and blockchain—offer robust security mechanisms for machine learning over networks, addressing data protection, privacy, and integrity in 6G environments.

In parallel, the convergence of quantum computing and machine learning has led to the emergence of quantum machine learning as a transformative approach to data analysis and decision-making. QML [4, 5] leverages the computational power of quantum computers to solve complex optimization and machine-learning tasks more efficiently than classical methods. Recent studies have investigated the application of QML techniques in various domains, including finance, healthcare, and cybersecurity. In the context of 6G networks, QML holds the potential to revolutionize data processing and analytics tasks at the network edge, enabling more intelligent and autonomous decision-making processes.

However, integrating QML with edge computing in 6G networks poses unique security, privacy, and trust challenges. As edge devices become increasingly interconnected and heterogeneous, ensuring the integrity and confidentiality of data transmitted over these networks becomes paramount. With its decentralized and immutable ledger, blockchain technology offers a promising solution to address these challenges. By providing a transparent and tamper-resistant record of transactions, blockchain enhances the security, privacy, and trust mechanisms in edge computing environments. Recent research has explored the application of blockchain in edge computing within 6G networks, focusing on its potential to enhance data provenance, access control, and auditability.

Furthermore, the theoretical underpinnings of quantum probability theory [6–8] and the central limit theorem are essential for understanding the probabilistic nature of quantum systems and its implications for statistical inference in QML tasks at the network edge. Quantum probability theory provides a framework for modeling and analyzing the behavior of quantum systems [9, 10], while the central limit theorem establishes the convergence of the distribution of sample means to a normal distribution, irrespective of the underlying distribution of the population. By incorporating insights from quantum probability theory and the central limit theorem, researchers can develop more robust and efficient QML algorithms tailored for edge computing environments within 6G networks.

Moreover, the proliferation of edge computing within 6G networks necessitates rethinking traditional security and privacy mechanisms. Centralized security approaches are no longer sufficient to protect against sophisticated threats targeting edge devices and infrastructure. Decentralized solutions, such as blockchain-based security mechanisms, offer a promising alternative by distributing trust and authority across the network. Recent advancements in blockchain technology, such as smart contracts and consensus mechanisms, enable secure and transparent data transactions in edge computing environments, fostering trust among network participants and mitigating the risk of malicious attacks.

3   Research Problem

This research endeavor aims to address critical challenges at the intersection of blockchain-based security and privacy, quantum machine learning (QML), and edge computing within the context of 6G networks. The overarching goal is establishing a secure and trusted ecosystem for collaborative QML tasks at the network edge while upholding data confidentiality and integrity.

Investigate the feasibility and effectiveness of integrating quantum-resistant cryptographic primitives into blockchain-based security mechanisms for edge computing environments within 6G networks, focusing on enhancing privacy and ensuring data confidentiality in collaborative quantum machine learning tasks. Investigate the impact of quantum computing advancements on the security and integrity of blockchain systems in edge computing within 6G networks, using Markovian models to analyze the dynamic evolution of quantum threats and their implications for privacy-preserving mechanisms. Evaluate the efficacy of machine learning techniques in predicting and mitigating quantum-related security risks to blockchain systems deployed in edge computing environments within 6G networks, focusing on enhancing security measures to protect quantum machine learning tasks. Assess the long-term sustainability of blockchain technology in the face of quantum computing advancements within 6G networks, considering the implications for data integrity, user trust, and overall system reliability in the context of edge computing and privacy preservation. Explore proactive solutions for safeguarding blockchain technology against quantum threats in edge computing scenarios within 6G networks, emphasizing the importance of timely research, development, and implementation of quantum-resistant cryptographic algorithms to uphold security and privacy in quantum machine learning tasks at the network edge. The research focuses on five key subproblems. Investigate the feasibility and effectiveness of integrating quantum-resistant cryptographic primitives into blockchain-based security mechanisms for edge computing environments within 6G networks, focusing on enhancing privacy and ensuring data confidentiality in collaborative quantum machine learning tasks. Investigate the impact of quantum computing advancements on the security and integrity of blockchain systems in the context of edge computing within 6G networks, using Markovian models to analyze the dynamic evolution of quantum threats and their implications for privacy-preserving mechanisms. Evaluate the efficacy of machine learning techniques in predicting and mitigating quantum-related security risks to blockchain systems deployed in edge computing environments within 6G networks, focusing on enhancing security measures to protect quantum machine learning tasks. Assess the long-term sustainability of blockchain technology in the face of quantum computing advancements within 6G networks, considering the implications for data integrity, user trust, and overall system reliability in the context of edge computing and privacy preservation. Explore proactive solutions for safeguarding blockchain technology against quantum threats in edge computing scenarios within 6G networks, emphasizing the importance of timely research, development, and implementation of quantum-resistant cryptographic algorithms to uphold security and privacy in quantum machine learning tasks at the network edge.

These research problems collectively aim to advance the understanding and development of robust security and privacy mechanisms, ensuring the integrity and reliability of edge computing applications within future wireless communication infrastructures. This research initiative addresses critical challenges arising at the nexus of blockchain-based security and privacy, quantum machine learning, and edge computing within 6G networks. The overarching objective is establishing a secure and trusted ecosystem for collaborative QML tasks at the network edge while safeguarding data confidentiality and integrity. Through five key subproblems, the research investigates integrating quantum-resistant cryptographic primitives into blockchain-based security mechanisms for edge computing environments, assesses the impact of quantum computing advancements on blockchain systems' security and integrity, evaluates machine learning techniques for predicting and mitigating quantum-related security risks, examines the long-term sustainability of blockchain technology in the face of quantum advancements, and explores proactive solutions for safeguarding blockchain against quantum threats in edge computing scenarios. These efforts collectively aim to advance the development of robust security and privacy mechanisms, ensuring the integrity and reliability of edge computing applications within future wireless communication infrastructures.

4   Model Architecture and Contribution

The model architecture proposed for securing and preserving privacy in edge computing environments within 6G networks encompasses several key components, each fulfilling specific functions to ensure the integrity and confidentiality of data processed at the network periphery. The architecture comprises the following elements:

Edge devices, such as sensors, IoT devices, and mobile devices, are located at the network's edge and are responsible for generating and transmitting data. These devices are primary data sources, collecting raw data from the environment or users, and often perform initial processing tasks like filtering and aggregation. Positioned close to the data source, edge devices reduce latency and optimize bandwidth by processing data locally before transmitting it. They execute edge computing tasks, enhancing privacy and security by minimizing the need to transmit sensitive data over the network. Edge devices are essential in various applications, including smart homes, healthcare, and industrial IoT, enabling real-time monitoring, data analysis, and decision-making.

Processing Time (T_{proc}): refers to the duration required for the AI agent to execute a specific task or set of tasks. This encompasses the computation time needed for data processing, algorithm execution, and decision-making within the agent. It can be quantified as follow.

$$T_{proc} = \sum_{i=1}^{n} t_i = T_{quantum} + T_{blockchain}$$

where, T_{proc} is the total processing time and t_i represents the individual processing times for each task i.

- Quantum Processing Time $(T_{quantum})$: Time taken for quantum computing tasks can be significantly faster for certain problems but also involve setup and readout times.
- Blockchain Processing Time $(T_{blockchain})$: Time required to validate and append transactions to the blockchain.

Transmission Time (T_{trans}): refers to the interval needed to transfer data between different components of the system, such as between the AI agent and a central server or between distributed ndoes.

- 6G Transmission Time (T_{6G}): Time taken to transmit data over a 6G network, which is expected to have extremely low latency and high bandwidth. Transmission time over a 6G network will depend on Data Size (D_i) of the amount of data to be transmitted, Bandwidth (B_{6G}) of the bandwidth of the 6G network, and Network Latency (L_{6G}) of the inherent latency in the 6G network, expected to be in the order of microseconds.

Additional Processing Time $(T_{additional})$:

- Additional Processing Time $(T_{additional})$: The additional processing time refers to the taken by an edge device or system to complete various computational tasks beyond the primary latency components, including quantum processing time, blockchain processing time, and 6G transmission time. The total latency (T_{total}) can be expressed as follows.

$$T_{total} = T_{quantum} + T_{blockchain} + T_{6G} + T_{additional}$$

Suppose we have the following data for a specific task i as follows.

- Total latency $(T_{total,i}) = 50ms$.
- Quantum processing time $(T_{quantum,i}) = 5ms$.
- Blockchain processing time $(T_{blockchain,i}) = 15ms$.
- Data size $(D_i) = 1000KB(8000Kbits)$.
- 6G network bandwidth $(B_{6G}) = 100Gbps(100,000,000\ Kbps)$.
- Network latency $(L_{6G}) = 1ms$.

Calculate the 6G transmission time as follows.

- $T_{6G,i} = \frac{D_i}{B_{6G}} + L_{6G} = \frac{8000Kbits}{100,000,000Kbps} + 1ms = 0.008ms + 1ms = 1.08ms$

The total latency equation as follows.

- $T_{total,i} = T_{proc,i} + T_{trans,i} + T_{addtional,i}$

- $T_{total,i} = T_{quantum,i} + T_{blockchain,i} + T_{6G,i} + T_{additional,i}$
- $50ms = 5ms + 15ms + 1.08ms + T_{additional,i}$
- $T_{additional,i} = 28.92ms$

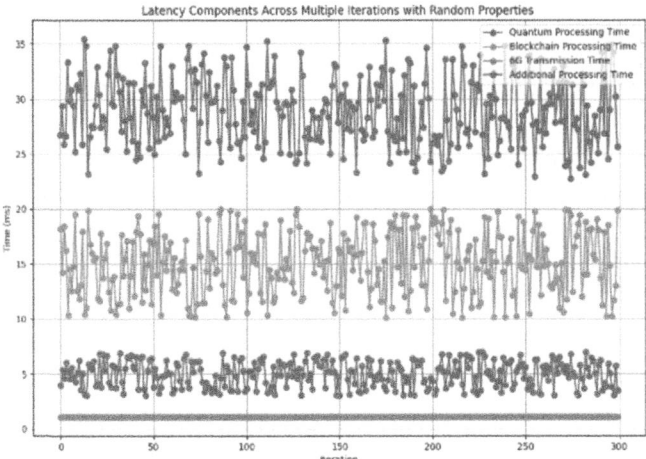

Fig. 2. Illustration of latency components in Quantum process time, Blockchain process time, 6G transmission time, and Additional process time across random properties.

The blockchain infrastructure serves as the underlying distributed ledger technology responsible for securing and immutably recording transactions and data interactions within the edge computing environment. Utilizing blockchain technology provides transparency, decentralization, and resilience against tampering or unauthorized access.

Quantum-resistant cryptographic algorithms are integrated into the blockchain infrastructure to protect data and transactions from potential threats posed by quantum computing advancements. These cryptographic primitives ensure that the security of the blockchain system remains intact even in the presence of powerful quantum adversaries.

Machine learning techniques, such as federated learning and differential privacy, enhance the security and privacy of collaborative QML tasks executed within the edge computing environment. These techniques enable data aggregation, model training, and inference while preserving the privacy of individual data contributors.

These modules consist of various security and privacy mechanisms, including access control, data encryption, and identity management, designed to enforce security policies and protect sensitive information within the edge computing environment. These modules work in conjunction with the blockchain infrastructure and quantum-resistant cryptography to ensure comprehensive security and privacy.

The orchestrator and governance layer oversees the coordination and management of activities within the edge computing environment. This layer governs access control policies, orchestrates data and task distribution among edge devices, and facilitates consensus mechanisms within the blockchain network.

Lastly, the orchestrator and governance layer oversee the coordination and management of activities within the edge computing environment, governing access control policies, orchestrating data and task distribution among edge devices, and facilitating consensus mechanisms within the blockchain network. Through the collective efforts of these components, the model architecture aims to establish a secure and privacy-preserving ecosystem for edge computing within 6G networks, paving the way for the seamless integration of emerging technologies into future wireless communication infrastructures.

As a contribution of this research, we introduce edge computing within 6G networks that promises to revolutionize data processing and communication by bringing computation closer to where data is generated. However, this paradigm shift also introduces complex security and privacy challenges, particularly in collaborative quantum machine learning tasks. To address these challenges, this research aims to explore integrating blockchain-based security mechanisms with quantum machine learning in edge computing environments [11–13]. By leveraging blockchain technology's decentralized and immutable ledger capabilities and quantum-resistant cryptographic primitives, the research endeavors to establish a robust framework that ensures data confidentiality, integrity, and privacy at the network periphery.

The research embarks on a multidimensional exploration, tackling five key subproblems. Firstly, it investigates the feasibility and effectiveness of integrating quantum-resistant cryptographic primitives into blockchain-based security mechanisms tailored for edge computing within 6G networks [14]. This investigation is crucial for enhancing privacy and data confidentiality in collaborative QML tasks. Secondly, the research delves into analyzing the impact of quantum computing advancements on the security and integrity of blockchain systems deployed in edge computing scenarios [15, 16]. Utilizing Markovian models, the dynamic evolution of quantum threats and their implications for privacy-preserving mechanisms are thoroughly assessed [9, 10]. Thirdly, the research evaluates the efficacy of machine learning techniques in predicting and mitigating quantum-related security risks to blockchain systems. This evaluation is essential for fortifying security measures and protecting QML tasks against potential threats.

Moreover, the study assesses the long-term sustainability of blockchain technology in the face of quantum advancements within 6G networks. Considerations for data integrity, user trust, and system reliability are carefully examined to ensure the continued viability of blockchain-based solutions in edge computing environments. Finally, the research explores proactive solutions for safeguarding blockchain against quantum threats, emphasizing timely research, development, and implementation of quantum-resistant cryptographic algorithms. These proactive measures are crucial for upholding security and privacy in future wireless communication infrastructures, paving the way for a secure and trusted ecosystem for edge computing.

5   Analysis of Simulation Results

Figure 3 illustrates the quantum graphic dynamics encompassing performance metrics and impact points, visually representing transitions between states 1, 2, and 3. In this context, performance refers to the system's ability to execute tasks efficiently and effectively, measured by various indicators such as processing power, latency, and error rates. The diagram provides a detailed visualization of how the system's state evolves over Time, highlighting the interactions and dependencies between various performance metrics and impact points.

State 1 represents the initial configuration of the quantum system, characterized by baseline performance metrics such as initial processing power, latency, and error rates. As the system transitions from state 1 to state 2, these metrics show a noticeable shift, indicating the system's response to external stimuli or internal adjustments. The visualization emphasizes the changes in key performance indicators, such as improved processing efficiency, reduced latency, and modified error rates, showcasing the dynamic nature of the quantum system.

State 2 serves as an intermediate phase where the system undergoes significant transformations. This state captures the real-time adjustments and optimizations within the quantum framework, illustrating how performance metrics adapt to achieve better outcomes. The impact points in this state highlight critical moments of change, such as the introduction of quantum error correction techniques or the implementation of advanced quantum algorithms. These impact points are crucial for understanding the system's adaptive capabilities and the effectiveness of various interventions.

The transition from state 2 to state 3 marks the final phase of the quantum dynamic process. In state 3, the system reaches a stable configuration with optimized performance metrics. The visualization in Fig. 3 underscores the culmination of previous transitions, demonstrating how initial challenges have been mitigated, and performance has been enhanced. The impact points in this state reflect the successful implementation of quantum enhancements, resulting in improved reliability, security, and overall system efficiency. Throughout the transitions between states 1, 2, and 3, the visualization in Fig. 3 provides a comprehensive overview of the quantum system's evolution. It effectively captures the interplay between different performance metrics and impact points, offering insights into the dynamic nature of quantum graphic dynamics and the continuous improvement process within the quantum computing framework.

In Fig. 4, the performance of nodes over time in state 1 is depicted through various lines representing different components. The blue line represents the baseline security and privacy (BS&P) measures, showcasing their stability and consistency throughout the observation period. The orange line represents the quantum machine learning (QML) component, highlighting its fluctuating performance over ime, influenced by external factors and system dynamics. The green line corresponds to edge computing capabilities, illustrating their gradual improvement over time in response to optimization efforts and technological advancements. The average performance metrics denoted by the dotted line at 75.46% are of particular interest. This average reflects a balanced level of performance across all components, indicating a satisfactory overall system performance. The associated impact value of 0.93 suggests a moderate influence on system behavior, indicating that changes in performance metrics have a discernible but not overwhelming effect on the system as a whole. Additionally, Fig. 4 showcases the best and worst-case scenarios for the QML component. The best-case scenario, represented by a performance level of 98.83%, demonstrates the peak capabilities of QML within the observed timeframe. This exceptional performance is accompanied by a high impact value of 9.66, indicating a significant influence on overall system behavior, potentially driving improvements in other components as well. Conversely, the worst-case scenario,

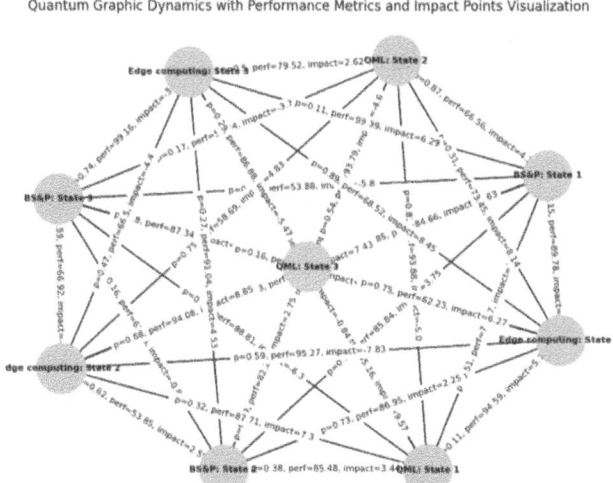

Fig. 3. Illustration of Quantum Graphic Dynamics.

with a performance level of 50.05%, represents the lowest observed performance for the QML component. Despite this setback, the associated impact value of 9.24 suggests that even suboptimal performance in QML can considerably impact the overall system dynamics, necessitating targeted interventions to mitigate adverse effects and restore performance to acceptable levels. In our analysis, 50.05% of the results achieved a specific performance level. This shows that half of the simulated scenarios met or exceeded the performance benchmarks. A performance level of 50.05% indicates that the framework is effective and performs well under these conditions. However, this also suggests that there is room for improvement. The results demonstrate the framework's capability to handle a substantial workload but also highlight areas for further optimization to enhance overall performance.

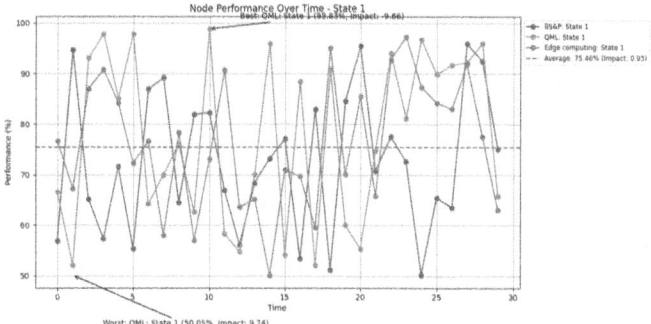

Fig. 4. Illustration of node performance over time in State 1.

Figure 5 illustrates the node performance over time for state 2, encompassing Blockchain Security and Privacy, Quantum Machine Learning, and Edge Computing components. While maintaining the same setup as state 1, state 2 shows a notable improvement with an average performance level of 77.60%, as denoted by the dotted line. This improvement reflects the ongoing optimization efforts across these technologies. However, it is crucial to note the variability within the Edge Computing component. In the best-case scenario, Edge Computing achieves a remarkable performance level of 99.47%, denoted by the upper bounds of its fluctuation, with an impact of -7.19.

Conversely, in the worst-case scenario, the performance drops to 50.08%, indicated by the lower bounds of its fluctuation, with an impact of -5.27. These extremes highlight the significant variability and potential impact of Edge Computing on overall node performance. Despite these fluctuations, the overall trend in Fig. 5 showcases a continuous and stable enhancement in node performance, emphasizing the cumulative benefits of optimization efforts across all components.

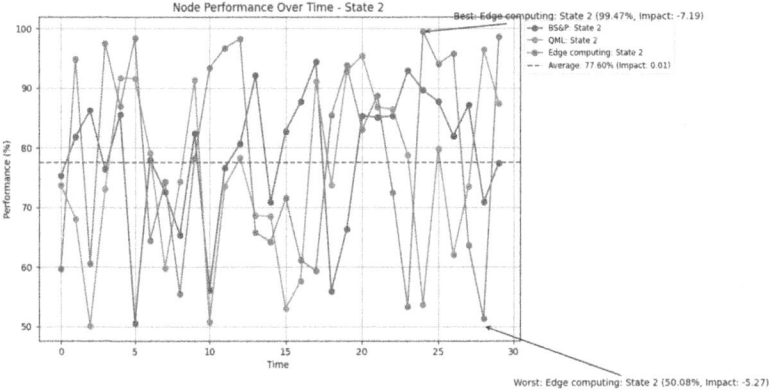

Fig. 5. Illustration of node performance over time in state 2.

Figure 6 depicts the node performance over time in state 3, where the average performance level reaches 77.84%, indicating a robust and stable system configuration. Within this state, notable variations are observed across different components. Quantum Machine Learning (QML) achieves its peak performance at 99.76%, which is marked as the highest point in the graph, with an impact of -1.6, signifying a significant positive influence on overall performance. Conversely, Blockchain Security and Privacy exhibit its lowest performance level in state 3, at 50.12%, representing the weakest link in the system. This decline is highlighted with an impact of 6.27, indicating substantial negative consequences. Despite the disparity between individual components, the system as a whole maintains a relatively high average performance, showcasing its resilience and adaptability. The visualization in Fig. 5 underscores the dynamic nature of node performance over time, emphasizing the importance of addressing vulnerabilities within specific components to ensure overall system efficacy and reliability.

The performance results in this study, derived from a meticulously designed simulation framework, offer a robust evaluation of the quantum system under various scenarios.

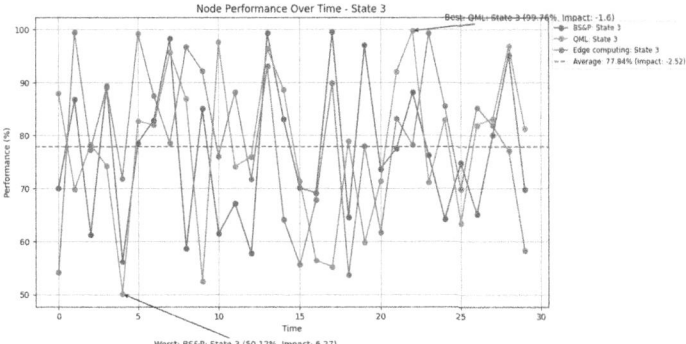

Fig. 6. Illustration of node performance over time in state 3.

This framework encompasses a detailed quantum system model, precise definitions and measurement methods for performance metrics, and clear representations of transition states and critical impact points. To ensure accuracy and reproducibility, random number generation is employed with specific parameters, including a defined seed value, and the use of uniform, normal, and exponential distributions tailored to different simulation aspects. Implemented in Python, the simulation leverages advanced tools and libraries such as 'networkx' for network modeling, 'matplotlib' for visualization, 'random' for stochastic processes, 'numpy' for numerical computations, and 'pandas' for data manipulation. The simulation runs multiple iterations to capture diverse scenarios comprehensively. The process includes initializing the model, executing simulations across states, and analyzing results to pinpoint trends and improvements. By detailing these implementation steps and parameters, this policy underscores the performance evaluation's transparency, credibility, and reliability, thereby providing a solid foundation for informed decision-making and future advancements in quantum system analysis.

6   Conclusion and Discussion

In conclusion, this paper presents a robust theoretical framework that integrates blockchain-based security and privacy mechanisms with quantum machine learning (QML) in the context of edge computing within 6G networks. This framework leverages blockchain's decentralized and immutable ledger to enhance access control, ensure data provenance, and verify integrity while incorporating quantum-resistant cryptographic primitives to guard against quantum threats. By integrating quantum probability theory, particularly through the central limit theorem, the framework effectively addresses the probabilistic nature of quantum systems and its implications for statistical inference in QML tasks.

When comparing the performance of the proposed framework with conventional approaches, several key differences and advantages emerge. Traditional methods often rely on centralized security models and classical machine learning algorithms, which can be less resilient to emerging threats and less efficient for complex computations. The blockchain-based security in our framework offers enhanced resilience against vulnerabilities and single points of failure, a significant improvement over centralized system.

Quantum computing integration in our framework demonstrates greater efficiency for complex tasks, with processing time growing more slowly compared to the faster but less effective classical algorithms used in conventional methods. Although blockchain processing time increases with transaction complexity, it provides superior security and data management compared to the quicker transaction processing of centralized systems. The high bandwidth of 6G networks ensures that our framework maintains minimal latency, unlike conventional approaches which may experience higher latency due to less optimized data handling. Additional processing tasks in our framework, such as data preprocessing and error checking, are crucial for maintaining system functionality and efficiency. In contrast, conventional methods may not address these aspects as thoroughly.

Visualizations in Figs. 4, 5, and 6 support this analysis by providing empirical evidence of the system's dynamics and performance characteristics, complementing the theoretical framework. Overall, the proposed framework represents a significant advancement over traditional methods, offering improved security, computational efficiency, and reliability. It lays a strong foundation for future research and development, advancing edge computing applications and paving the way for more secure and efficient wireless communication infrastructures.

References

1. Giordani, M., Polese, M., Mezzavilla, M., Rangan, S., Zorzi, M.: Toward 6G networks: use cases and technologies. IEEE Commun. Mag. **58**(3), 55–61 (2020). https://doi.org/10.1109/MCOM.001.1900411
2. Letaief, K.B., Chen, W., Shi, Y., Zhang, J., Zhang, Y.-J.A.: The roadmap to 6G: AI empowered wireless networks. IEEE Commun. Mag. **57**(8), 84–90 (2019). https://doi.org/10.1109/MCOM.2019.1900271
3. Nawaz, S.J., Sharma, S.K., Wyne, S., Patwary, M.N., Asaduzzaman, M.: Quantum machine learning for 6G communication networks: state-of-the-art and vision for the future. IEEE Access **7**, 46317–46350 (2019). https://doi.org/10.1109/ACCESS.2019.2909490
4. Caro, M.C., et al.: Generalization in quantum machine learning from few training data. Nat. Commun. **13**(1), 4919 (2022)
5. Khan, T.M., Robles-Kelly, A.: Machine learning: quantum vs classical. IEEE Access **8**, 219275–219294 (2020)
6. Gudder, S.P.: Quantum probability. Academic Press (2014)
7. Accardi, L.: Topics in quantum probability. Phys. Rep. **77**(3), 169–192 (1981)
8. Davies, E.B., Lewis, J.T.: An operational approach to quantum probability. Commun. Math. Phys. **17**(3), 239–260 (1970)
9. Seol, J., Kim, J.: Machine learning ensures quantum-safe blockchain availability. J. Comput. Inf. Syst. 1–25 (2024)
10. Seol, J., Kim, J.: Quantum threat and dependability of quantum-safe blockchain-based distributed control systems and network. Issues Inf. Syst. **24**(3) (2023)
11. Tanwar, S., Bhatia, Q., Patel, P., Kumari, A., Singh, P.K., Hong, W.C.: Machine learning adoption in blockchain-based smart applications: the challenges, and a way forward. IEEE Access **8**, 474–488 (2019)
12. Gill, S.S.: Quantum and blockchain based serverless edge computing: a vision, model, new trends and future directions. Internet Technol. Lett. **7**(1), e275 (2024)

13. Cherbal, S., Zier, A., Hebal, S., Louail, L., Annane, B.: Security in internet of things: a review on approaches based on blockchain, machine learning, cryptography, and quantum computing. J. Supercomput. **80**(3), 3738–3816 (2024)
14. Kim, M., Oh, I., Yim, K., Sahlabadi, M., Shukur, Z.: Security of 6G enabled vehicle-to-everything communication in emerging federated learning and blockchain technologies. IEEE Access (2023)
15. Guo, S., Hu, X., Guo, S., Qiu, X., Qi, F.: Blockchain meets edge computing: a distributed and trusted authentication system. IEEE Trans. Industr. Inf. **16**(3), 1972–1983 (2019)
16. Yang, R., Yu, F.R., Si, P., Yang, Z., Zhang, Y.: Integrated blockchain and edge computing systems: A survey, some research issues and challenges. IEEE Commun. Surv. Tutorials **21**(2), 1508–1532 (2019)
17. Shi, W., Cao, J., Zhang, Q., Li, Y., Xu, L.: Edge computing: vision and challenges. IEEE Internet Things J. **3**(5), 637–646 (2016)
18. Al-Ansi, A., Al-Ansi, A.M., Muthanna, A., Elgendy, I.A., Koucheryavy, A.: Survey on intelligence edge computing in 6G: characteristics, challenges, potential use cases, and market drivers. Future Internet **13**(5), 118 (2021)
19. Liao, Z., et al.: Distributed probabilistic offloading in edge computing for 6G-enabled massive internet of things. IEEE Internet Things J. **8**(7), 5298–5308 (2020)
20. Nawaz, S.J., Sharma, S.K., Wyne, S., Patwary, M.N., Asaduzzaman, M.: Quantum machine learning for 6G communication networks: state-of-the-art and vision for the future. IEEE access **7**, 46317–46350 (2019)
21. Siriwardhana, Y., Porambage, P., Liyanage, M., Ylianttila, M.: AI and 6G security: opportunities and challenges. In: 2021 Joint European Conference on Networks and Communications and 6G Summit (EuCNC/6G Summit) IEEE, pp. 616–621 (2021)

An Analysis of Malicious Behaviors of Open-Source Packages Using Dynamic Analysis

Thanh-Cong Nguyen[1], Duc-Ly Vu[2(✉)], and Narayan C. Debnath[2]

[1] University of Information Technology, Ho Chi Minh City, Vietnam
20521143@gm.uit.edu.vn
[2] School of Computing and Information Technology, Eastern International
University, Binh Duong, Vietnam
{ly.vu,narayan.debnath}@eiu.edu.vn

Abstract. There has been an increasing number of malicious open-source packages in recent years. A recent backdoor attack on the Linux *xz* utility has highlighted the importance of security checks on open-source packages, especially popular ones. While major security scanners focus on identifying vulnerabilities (CVEs) in open-source packages, there are very few studies on malware analysis techniques for them. Similar to traditional malware analysis, there are two types of analysis for open-source packages: static and dynamic analysis. Static analysis techniques mainly focus on analyzing the source code of a package while dynamic analysis techniques execute the code in an isolated environment. Dynamic analysis techniques seem more promising than static analysis techniques, as they can expose packages' behaviors at runtime. However, current dynamic analysis tools (e.g., *package-analysis*) make minimal effort to provide insight into the behaviors of open-source packages. In this paper, we attempt to analyze the dynamic behaviors of open-source packages on popular package repositories, including npm, PyPI, RubyGems, Packagist, and crates.io. We also analyze the discrepancies in behaviors between benign and malicious packages at runtime, which is helpful in building rules for malware detection. Our study finds that malicious packages perform a significantly higher number of domain communications and command executions. Malicious packages use simple techniques for malicious operations such as *base64* or *curl*.

Keywords: Dynamic malware analysis · Open-source malicious packages · Open-source software security · Software supply chain Security · Software supply chain attacks

1 Introduction

In modern software development, developers usually use third-party open-source packages or libraries from language-based package repositories (e.g., PyPI for Python). This practice saves developers' time and improves development velocity. Besides the benefits of using open-source packages, there are security risks in package repositories. For example, attackers could implant malicious code in

G. Hu et al. (Eds.): CAINE 2024, CCIS 2242, pp. 102–114, 2025.
https://doi.org/10.1007/978-3-031-76273-4_8

the source code repository of a popular package to infect its users. The recent *xz* attack magnifies the importance of scanning open-source code before using it [8].

Researchers and commercial organizations have proposed various techniques and developed tools for scanning malicious packages. Basically, the tools can be categorized into: static and dynamic malware analysis tools. Static analysis tools only examine package information (e.g., source code or metadata) without executing them while dynamic analysis performs package execution in an isolated environment. While static analysis tools are fast and easy to implement, they seem to be ineffective against anti-analysis techniques such as code obfuscation [15]. In addition, statically analyzing packages could introduce many false positives [30], one of the static malware detection tools called *OSSGadget* [14] even highlights this limitation on their GitHub page.

Dynamic malware analysis techniques, on the other hand, run the code in an isolated environment, typically a sandbox, and observe its behaviors, such as system calls or network connections. Current dynamic analysis tools for open-source packages are promising but still seem to be immature [30]. For example, *package-analysis*, a dynamic analysis tool developed by *OpenSSF*, has been used to detect malicious packages [18]. However, we notice that the tool only provides raw analysis results in JSON format, which requires extensive analytical effort. This effort involves either looking at the raw results or writing detection rules to decide whether a package is malicious. In this regard, this paper takes a step forward in mining the raw outputs of *package-analysis* and providing insights into the behaviors of benign and malicious packages in popular package repositories.

When analyzing open-source packages, researchers typically look for malicious indicators (e.g., suspicious domains or system calls) in the analysis report based on their expert knowledge to decide whether a sample is malicious. False positives can occur in the analysis, where a benign package is mistakenly flagged as malicious [30]. To prevent false positives, malware detection tools need to differentiate malicious behaviors from benign ones. To the best of our knowledge, there is no study in the literature analyzing the malicious behaviors of open-source packages using dynamic analysis.

This paper studies the behaviors and characteristics of benign and malicious open-source packages in popular package repositories, including crates.io, npm, Packagist, PyPI, and Rubygems. We have created a dataset consisting of malicious and benign packages. We then analyze the dataset to understand the differences between benign and malicious behaviors. Our analysis shows that the malicious packages perform a significantly higher number of domain communications and command executions compared to benign packages. The malicious packages, however, employ simple techniques such as *base64* for data encoding and curl for exfiltrating users' information to a remote server.

In summary, this paper has the following contributions:

- A methodology for curating malicious and benign packages
- An investigation of a dynamic malware analysis tool called *package-analysis* for open-source packages

– An analysis of malicious behaviors and benign behaviors in open-source packages

2   Background

2.1   Software Supply Chain Attacks

Software supply chain attacks occur when attackers inject malicious code into a component in the software supply chain [26]. End users can be infected by downloading or updating the software product. Ladisa et al. [13] present a taxonomy of software supply chain attacks on package managers and their countermeasures. In their paper, typosquatting and combosquatting techniques are the most popular methods to confuse end users and lure them into downloading malicious packages. Several approaches [24,31] have been proposed to detect this type of attack.

One notable example of the software supply chain attack is the *SolarWinds* attack where attackers were able to inject malicious code into a company software's update [5]. Recently, malicious code was discovered in the upstream tarballs of *xz*, starting with version 5.6.0. The tarballs included extra .m4 files, which contained instructions for building with automake that did not exist in the repository. These instructions, through a series of complex obfuscations, extract a prebuilt object file from one of the test archives, which is then used to modify specific functions in the code while building the *liblzma* package. This issue results in *liblzma* being used by additional software, such as *sshd*, to provide functionality that will be interpreted by the modified functions [9].

2.2   Static Malware Analysis Tools

Static malware analysis techniques look for malicious patterns in the source code or metadata of a package. Although these techniques are lightweight, they cannot catch malicious code that is only executed at runtime. Moreover, static analyzers are prone to anti-analysis techniques, such as code obfuscation. There are existing static malware scanners, including the following:

– OSS Detect Backdoor [14]: an open-source tool developed by Microsoft. It contains a set of utilities for investigating different aspects of an open-source package.
– Bandit4Mal [28]: a tool developed by academic researchers at the University of Trento and SAP Security Research. This tool scans Python packages for malicious traits using AST analysis and hand-written malware detection rules [29].
– PyPI Malware Checks [19]: a tool used by PyPI to check for suspicious code lines in every package uploaded to PyPI. The tool uses a set of regular expression-based rules.
– Capslock [12]: a capability analysis command line interface (CLI) program for Go packages that informs users of which privileged operations a given package can access. Currently, Capslock only supports Go.

The above-mentioned tools normally parse the code of a package into ASTs and use a set of rules to look for malicious patterns. The study by Vu et al. [30] evaluates different static malware detection tools for open-source packages and reports that they generate many false positives. The same study suggests considering dynamic scanning techniques, in particular running code in a sandbox, for better malware detection.

2.3   Dynamic Malware Analysis Tools

Dynamic analysis tools require executing the source code of a package in an isolated environment. By running the code, these tools record execution traces of the package, such as running processes, executed commands, communicated IPs/domains, and accessed files. Although dynamic analysis tools can provide a more precise analysis, they usually take time and require setting up an appropriate environment. The following are dynamic analysis tools:

- MalOSS [10]: a tool that uses Sysdig [23] as the tracing tool to capture system call traces related to IPs, DNS queries, files, and processes.
- package-analysis [17]: an open-source dynamic analysis tool, developed by Google in 2022. The tool monitors command, file, and network activities in a sandbox called *Gvisor* (discussed later in this paper).
- package-hunter [11]: a tool that analyzes a program's dependencies for malicious code. The tool installs the dependencies in a sandbox environment and monitors system calls executed during the installation [6].

In this paper, we choose *package-analysis* as our analysis tool because it is an open-source tool. The tool analyzes the capabilities of packages available on open-source repositories. The tool looks for behaviors that indicate malicious software: 1) *What files do they access?*, 2) *What addresses do they connect to?*, and 3) *What commands do they run?* [17]. *package-analysis* captures malicious interactions with the system as well as network connections that can be used to leak sensitive data or allow remote access using the *Gvisor* sandbox [7]. In addition, the raw outputs of *package-analysis* are available on Google BigQuery [2] enabling us to perform an in-depth analysis of the behaviors of benign and malicious packages.

3   Sandboxing

A sandbox is an isolated environment used to dynamically execute suspicious code. In this way, we can run untrusted programs in a safe environment without affecting real systems [22]. In this section, we discuss the architecture and limitations of *Gvisor*, the sandbox that underpins *package-analysis*.

Figure 1 shows the overall architecture of *package-analysis*. This architecture includes a nested container, where the sandbox uses a container inside another container. By doing so, *package-analysis* creates a safe, isolated environment for executing suspicious code. In particular, the outer container uses a Docker image called *gcr.io/ossf-malware-analysis/analysis* to create the inner container. *Gvisor* runs on multiple architectures (as Linux does) such as x86, ARM, and Virtual Machines (VMs). *Gvisor* intercepts all system calls from sandboxed applications to the Linux kernel. Although *Gvisor* provides beneficial features, it has the following limitations:

- *Gvisor* runs in user space, so it has a lower priority than the kernel when executed.
- Currently, *Gvisor* does not provide a complete list of system calls like the kernel. In particular, *Gvisor* only supports the 211 most common system calls. For unsupported system calls, *Gvisor* will not process them and will raise exceptions.
- Applications in *Gvisor* cannot interact with the hardware of the host machine, as *Gvisor* creates a protected layer to prevent any interaction between applications and the host machine [25].

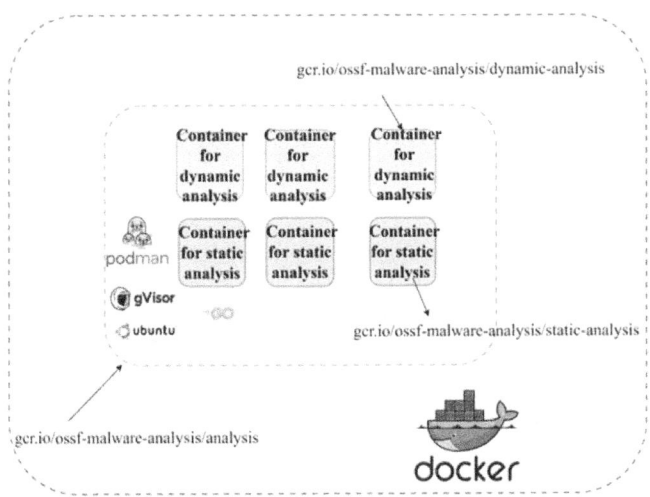

Fig. 1. Package Analysis Sandbox Architecture.

4 Data Collection

This section presents our data collection and analysis workflow. In particular, we present two datasets: malicious packages and benign packages.

4.1 Malicious Packages Collection

Figure 2 shows our data collection workflow. We collected malicious open-source packages from the following sources:

– Vulert [4]: this service provides security information (such as CVE IDs) about open-source packages in popular package repositories such as npm, PyPI, RubyGems, crates.io.
– Vulners [1]: Vulners maintains a database of software vulnerabilities. Vulner also provides Application Programming Interfaces (APIs) to search for specific software vulnerabilities.
– OSV [3] monitors open-source packages for vulnerabilities. Like Vulners, this service provides APIs to query security information about open-source packages, including the identification of malicious packages.

After obtaining malicious package names and their descriptions (e.g., behavior descriptions), we query the analysis results of the malicious packages on Google BigQuery's *ossf-package-analysis* [2]. Next, we extract the executed commands, IP addresses, and domains for each package analysis report for further analysis (as shown in Sect. 5).

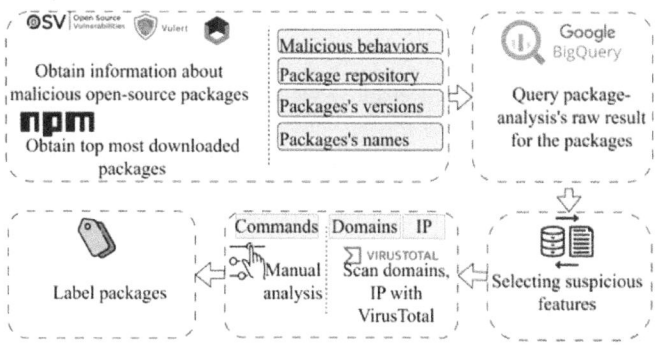

Fig. 2. Our data collection and analysis workflow.

Table 1 shows the statistics of the malicious packages we have collected. On average, each package has two versions. We notice that npm has the most packages and versions (approximately 90%) in the collected dataset. Conversely, crates.io has the fewest packages and versions. This might indicate that, at the moment, npm is the most valuable target for attackers to inject malicious code. Researchers might need to pay more attention to vetting this repository for safer use.

4.2   Benign Packages Collection

Following Zahan et al. [32] and Vu et al. [30], we collected the top 1000 most downloaded open-source packages on npm as benign packages. Note that we chose npm packages because they are the most common packages in the malicious packages dataset. In the end, we have a balanced dataset that has a fair number of packages. We then queried the raw analysis for those benign packages on

Table 1. Statistics of collected malicious open-source packages.

	#Packages	#Versions
crates.io	1	10
npm	1041	2293
PyPI	113	216
RubyGems	16	27
Total	1171	2546

Google BigQuery [2]. The raw results of the benign packages are analyzed and compared with those of the malicious packages in the next section.

5   Findings

5.1   Performance of Package-Analysis on Open-Source Packages

In this section, we investigate the dataset published by OSSF on Google Big-Query called *ossf-malware-analysis* [2]. This dataset contains the live analysis of *package-analysis* on open-source packages from crates.io, npm, PyPI, Packagist, RubyGems package repositories. Table 2 summarizes the analysis performance of *package-analysis* on the packages from the supported package repositories. We can see that *package-analysis* covers most of the packages on crates.io (87.2%) while having the lowest coverage (16.37%) for Packagist. Surprisingly, although npm has the highest number of packages, only about 28% of its packages have been analyzed by *package-analysis*.

Table 2. Statistics of open-source packages on Package Analysis's BigQuery.

Repository	Language	#Packages in repository	#Packages analyzed by package-analysis	Ratio of analyzed packages and total packages
crates.io	Rust	144 047	125 640	87.22%
npm	Javascript	4 530 434	1 264 900	27.92%
Packagist	PHP	390 942	63 987	16.37%
PyPI	Python	535 457	287 299	53.65%
RubyGems	Ruby	197 071	31 803	16.14%

Figure 3 shows the completion rates of importing and installing packages in different repositories. We observed that on average, *package-analysis* has success rates of 62.75% and 95.81% when installing and importing a package, respectively. Packages in npm and crates.io have the highest success rate when being installed and imported, respectively. However, crates.io has the lowest success rate when being installed by *package-analysis*. This indicates that installing a Rust package in crates.io is still a challenging problem.

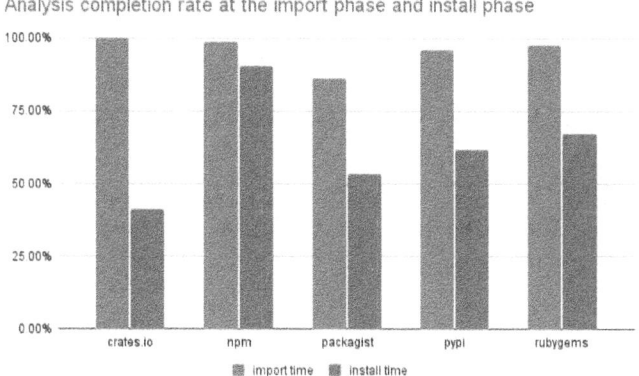

Fig. 3. Analysis completion rate of package-analysis at the import phase and install phase.

5.2   Analysis of Malicious and Benign Packages

In this section, we report our findings on the behaviors exposed by the benign and malicious packages in our collected dataset. Table 3 shows the behaviors of malicious packages in crates.io, npm, PyPI, and RubyGems. Note that we could not find records for the packages in Packagist from our sources in Subsect. 4.1. Table 3 shows that at least one malicious package from each package repository communicates with a domain associated with malicious activity. The packages in npm and PyPI also execute one or more commands associated with malicious behavior. One-third of the npm packages exhibit both domain communication and command execution behaviors.

Table 3. Frequency Statistics of Suspicious behaviors in Benign and Malicious Datasets.

	Communicates with a domain associated with malicious activity	Executes one or more commands associated with malicious behavior	Communicates with a domain associated with malicious activity and executes one or more commands associated with malicious behavior.
crates.io	1	0	0
npm	614	106	321
PyPI	86	5	22
RubyGems	16	0	0

Malicious packages can communicate with a malicious domain to download additional malware, known as *droppers*. Several reports have shown that npm and PyPI packages have been found to install Linux cryptominers, information stealers, or Windows Trojans [20, 21, 27]. Malicious packages can download a script from a C&C server and then execute it on the victim's system.

Table 4. Frequency Statistics of Malicious Indicators in Benign and Malicious Datasets.

Dataset	#Commands	#Unique Commands	#URLs	#Unique URLs	#IP Addresses	#Unique IP Addresses
Malicious	2 845 356	533	68 677 299	934	74 584 405	682
Benign	1 310 054	818	5 024 881	7	5 007 963	134
Total	4 155 410	2760	73 702 180	941	79 592 368	816

Commands Analysis. Malicious packages frequently run system commands on the victim's systems. For example, malicious packages can use the *base64* command to encode users' information before sending it out to a remote server. Table 4 shows the occurrences of commands, IP addresses, and URLs in the benign, malicious, and combined sets. We observe that malicious packages execute twice as many commands as benign packages. However, malicious packages often perform the same commands multiple times, which might indicate that malicious actors could reuse the code.

Table 5 shows the top ten commands executed by the malicious packages in our collected dataset. It is evident that most commands relate to information-gathering activities. It is noted the malicious packages employ simple techniques for malicious tasks, such as using *base64* for data encoding. Compared to traditional malware on Windows or Linux, malicious packages in the package repositories are much simpler, and some are distributed as Proof-of-Concepts (POCs). However, it is expected that malicious packages will evolve in both quantity and quality.

Compared to the top commands in the malicious packages, *ls* remains the most common command executed by the benign packages. In the benign packages, *grep* is the second most common command, used for searching and manipulating text patterns within files. Similar to the malicious packages, the benign packages also run some commands such as *uname* to obtain system information, including the name of the operating system. However, the benign packages do not frequently execute shell-related commands, such as *bash*. The *bash* command starts a new shell within the original shell, allowing attackers to run additional commands or shellcodes.

Domains and IP Addresses Analysis. Malicious packages typically communicate with external servers, commonly referred to as Command and Control (C&C) servers, to receive instructions or send stolen data. In this section, we analyze the domains contacted by malicious packages for communication purposes. Figure 5 shows that the majority of the domains are flagged by a security

Table 5. Top ten commands executed by the malicious packages in our dataset.

Command	#Occurences	Description	Malicious behavior
ls	87 706	Lists computer files and directories	Information gathering
bash	87 656	Starts a new bash shell	Command execution
cat	87 500	Views the contents of a file	Information gathering
dpkg-query	82 758	Shows information about dpkg packages.	Information gathering
lsb_release -a	82 758	Gets distribution-specific information.	Information gathering
base64	78 146	Encodes and Decodes data	Data hiding
/usr/bin/curl	77 026	Tranfers data using various network protocols.	Data infiltration
which	68 900	Identifies the location of executables	Information gathering
which bash	68 652	Identifies the location of the bash executable	Information gathering
tr	64 524	Translates or Deletes characters	Data hiding

vendor. At least 50 domains are flagged by two or more security vendors in Virus-Total. The higher the number of flags raised by security vendors for a domain, the greater the confidence that the domain is malicious.

Table 4 shows that the malicious packages communicate with significantly more domains (URLs) than benign packages - nearly 14 times as many. Furthermore, there is a much greater diversity of domains contacted by malicious packages - nearly 133 times more than those contacted by benign packages. This observation might indicate that malicious actors originate from different groups or that they frequently change their C&C servers after being detected.

Interestingly, it appears that malicious packages appear to be using Out-of-band Application Security Testing (OAST) tools when probing for Common Vulnerabilities and Exposures (CVEs). We identified the following OAST domains used in probing attempts for CVEs. Attackers may have scanned victims to find vulnerable targets. Additionally, these domains include exploits that have led attackers to install cryptominers on compromised hosts [16]. Table 6 presents the OAST domains associated with the malicious packages in our dataset. All the domains are flagged by at least two security vendors in VirusTotal. Most of the domains were flagged as malicious, malware, or phishing by the security vendors.

Besides domain names, IP addresses are among the basic features of network activity. IP addresses embedded in malicious packages usually indicate the locations of command and control servers. However, in benign packages, IP addresses may represent the locations of database servers or other legitimate services. Table 4 shows that there are nearly 15 times more IP addresses in malicious packages than in benign packages. Out of 37 423 IP addresses, 15 927 IP addresses (42.56%) were recognized as malicious by at least one security vendor in VirusTotal. Figure 4 shows that most of the IP addresses in the benign packages are in the United States. We observe that Germany has the second-largest number of IP addresses, which could suggest that malicious packages targeted users in Europe.

Table 6. Out-of-band Application Security Testing (OAST) domains utilized by malicious packages during probing attempts for Common Vulnerabilities and Exposures (CVEs).

OAST Domain	#Flagged AVs	Labels assigned by AVs
oast.fun	11	Malicious, Suspicious, Phishing
oast.me	11	Malicious, Malware, Phishing
oast.live	10	Malicious, Malware, Phishing
oast.pro	10	Malicious, Malware, Phishing, Suspicious
oast.site	10	Malicious, Malware, Phishing, Suspicious
oast.online	7	Not Recommended, Malicious, Phishing, Suspicious
oastify.com	2	Malicious

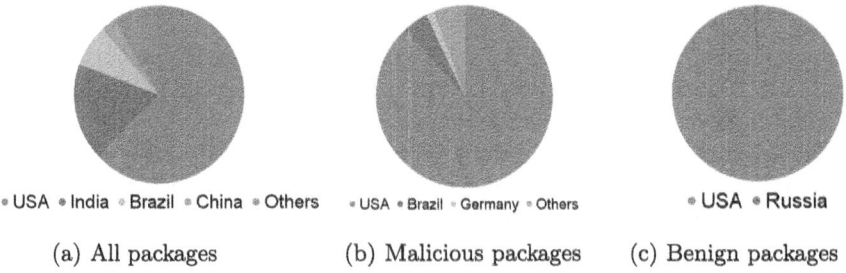

  • USA • India • Brazil • China • Others   • USA • Brazil • Germany • Others   • USA • Russia

(a) All packages   (b) Malicious packages   (c) Benign packages

Fig. 4. Geographic locations of IP Addresses found in open-source packages.

6   Limitations and Future Work

Currently, we have only analyzed the behaviors of open-source packages in the Linux environment because *package-analysis* currently supports only a Linux sandbox. Our next plan is to extend its capability to support the Windows environment. This will require developing a Windows kernel and its utilities. In addition, engineering efforts will be necessary to improve the analysis completion rates of *package-analysis* (as shown in Fig. 3), particularly in the installation phase.

In our study, we observe that *package-analysis* does not perform well during the installation phase (as shown in Fig. 3). This limitation may prevent our analysis from assessing the behaviors of all the packages in the studied repositories. Therefore, our next step is to investigate the *package-analysis* logs to identify and resolve the errors.

Our analysis, based on *package-analysis*, does not indicate whether a package is malicious. Users of the tool must manually analyze the raw results it produces to make informed decisions. A promising direction is to apply machine learning techniques to the raw analysis generated by *package-analysis* on Google Big-Query [2]. For example, commands, domain URLs, and IP addresses could serve as valuable features for machine-learning-based malware detection approaches.

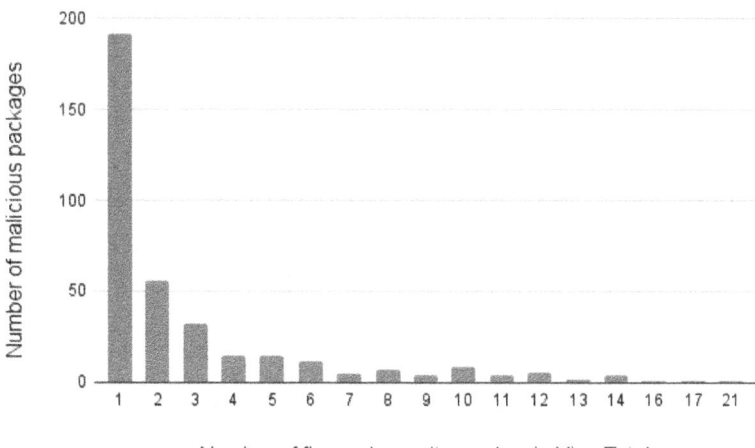

Number of flagged security vendors in VirusTotal

Fig. 5. Distribution of number of domains flagged by security vendors in VirusTotal

7 Conclusion

In this paper, we have taken a closer look at a dynamic analysis tool called *package-analysis*, including its sandbox technique and results. We have analyzed the raw results of *package-analysis* for open-source packages in popular repositories to identify common malicious behaviors. Our analysis indicates that malicious packages employ simple techniques such as *base64* for data encoding or *curl* for data transfer. Compared to benign packages, malicious packages exhibit significantly higher activity in command executions and domain communications.

In summary, this paper contributes to the analysis of open-source malware through dynamic analysis. This contribution could benefit security researchers in analyzing malicious open-source packages and developing detection tools.

References

1. CVE Database - Security Vulnerabilities and Exploits
2. Google BigQuery's OSSF-malware-analysis
3. OSV - Open Source Vulnerabilities
4. Vulert: Software Composition Analysis & Vulnerability Alerts
5. Alkhadra, R., Abuzaid, J., AlShammari, M., Mohammad, N.: Solar winds hack: In-depth analysis and countermeasures. In: 2021 12th International Conference on Computing Communication and Networking Technologies (ICCCNT), pp. 1–7. IEEE (2021)
6. Appelt, D.: Meet package hunter: A tool for detecting malicious code in your dependencies (2021)
7. gVisor Authors, T.: The container security platform
8. Bals, F.: What is the XZ utils backdoor : Everything you need to know about the supply chain attack (2024)

9. Database, G.A.: Malicious code was discovered in the upstream tarballs of... (2024)
10. Duan, R., Alrawi, O., Kasturi, R.P., Elder, R., Saltaformaggio, B., Lee, W.: Towards measuring supply chain attacks on package managers for interpreted languages. arXiv preprint arXiv:2002.01139 (2020)
11. GitLab: Package hunter: A tool for identifying malicious dependencies via runtime monitoring (2020)
12. Google: Github - google/capslock: A capability analysis cli for go packages (2020)
13. Ladisa, P., Plate, H., Martinez, M., Barais, O.: SoK: Taxonomy of attacks on open-source software supply chains. In: 2023 IEEE Symposium on Security and Privacy (SP), pp. 1509–1526. IEEE (2023)
14. Microsoft: OSS detect backdoor (2019)
15. Moser, A., Kruegel, C., Kirda, E.: Limits of static analysis for malware detection. In: Twenty-Third Annual Computer Security Applications Conference (ACSAC 2007), pp. 421–430. IEEE (2007)
16. Networks, U..P.A.: Threat brief: Multiple ivanti vulnerabilities (2024)
17. Ossf: Github - ossf/package-analysis: Open source package analysis (2022)
18. OSSF: Package analysis: Case studies (2022)
19. PyPA: Malware checks (2020)
20. Sharma, A.: 241 npm and PyPI packages caught dropping linux cryptominers (2022)
21. Sharma, A.: Attacker floods PyPI with 1000s of malicious packages that drop windows trojan via dropbox (2023)
22. Sikorski, M., Honig, A.: Practical malware analysis: the hands-on guide to dissecting malicious software. no starch press (2012)
23. Sysdig: Security for containers, kubernetes, and cloud
24. Taylor, M., Vaidya, R., Davidson, D., De Carli, L., Rastogi, V.: Defending against package typosquatting. In: Network and System Security: 14th International Conference, NSS 2020, Melbourne, VIC, Australia, November 25–27, 2020, Proceedings 14, pp. 112–131. Springer (2020)
25. Tech, G.C.: Sandboxing your containers with gvisor (cloud next '18) (2018)
26. Tessian: What is a software supply chain attack? (2023)
27. thehackernews: Malicious PyPI packages slip whitesnake InfoStealer malware onto windows machines (2024)
28. Vu, D.L.: A fork of bandit tool with patterns to identifying malicious python code (2020)
29. Vu, D.L., Massacci, F., Pashchenko, I., Plate, H., Sabetta, A.: LastPyMile: identifying the discrepancy between sources and packages. In: Proceedings of the 29th ACM Joint Meeting on European Software Engineering Conference and Symposium on the Foundations of Software Engineering, pp. 780–792 (2021)
30. Vu, D.L., Newman, Z., Meyers, J.S.: Bad snakes: Understanding and improving python package index malware scanning. In: 2023 IEEE/ACM 45th International Conference on Software Engineering (ICSE), pp. 499–511. IEEE (2023)
31. Vu, D.L., Pashchenko, I., Massacci, F., Plate, H., Sabetta, A.: Typosquatting and combosquatting attacks on the python ecosystem. In: 2020 IEEE European Symposium on Security and Privacy Workshops (EuroS&PW), pp. 509–514. IEEE (2020)
32. Zahan, N., Zimmermann, T., Godefroid, P., Murphy, B., Maddila, C., Williams, L.: What are weak links in the npm supply chain? In: 2022 IEEE/ACM 44th International Conference on Software Engineering: Software Engineering in Practice (ICSE-SEIP), pp. 331–340. IEEE (2022)

Accessible Cybersecurity Education Using Prompt Tree

Pavan Subhash Chandrabose Nara[1], Reshmi Mitra[1(✉)], Indranil Roy[1],
and T. Robin Cole III[2]

[1] Southeast Missouri State University, Cape Girardeau, USA
{pnara1s,rmitra,iroy}@semo.edu
[2] Rite Group, Cape Girardeau, USA
trcole3@theritegroup.com

Abstract. This paper presents a novel approach to developing an innovative prompt tree structure for cybersecurity education and training. It aims to address the challenges of accessibility and technical complexity within the field. The proposed framework utilizes customizable prompts and recipe-style interactions with Large Language Models (LLMs) to guide users in addressing cybersecurity issues. This framework is inspired by the popular fishbone diagram and utilizes credible cybersecurity frameworks such as OWASP and NVD. The system is designed to empower individuals, regardless of their expertise, to effectively understand and implement cybersecurity measures.

1 Introduction

The escalating volume and sophistication of cyberattacks pose an ever-growing threat to individuals, businesses, and critical infrastructure, with the projected annual cost of cybercrime reaching $10.5 trillion by 2025 [11]. As the threat landscape evolves, artificial intelligence (AI), particularly Large Language Models (LLMs), is emerging as a potent countermeasure [7]. LLMs are being integrated directly into software tools, such as GitHub's Co-Pilot [2,14,23] and integrated development environments (IDEs) like IntelliJ [16] and Visual Studio Code, enabling software teams to access these tools seamlessly [7]. With 82% of organizations recognizing potential of LLMs for enhanced breach detection and response, the impact of AI on cybersecurity is clear. Companies that leverage AI tools are able to detect breaches significantly faster, underscoring the critical role of AI in fortifying cybersecurity defenses.

However, a fundamental obstacle remains: the acute shortage of *skilled cybersecurity professionals*. Traditional cybersecurity training methods struggle to keep pace with the rapidly evolving vulnerabilities and attack vectors, while the prevalence of specialized tools requires substantial technical expertise, limiting access for non-technical users. This creates a significant gap between the urgent need for cybersecurity knowledge and the accessibility of effective training resources. Large Language Models (LLMs) [3] offer a compelling opportunity to bridge this gap. Their ability to process vast amounts of information, generate human-quality text, and engage in conversational interactions holds the

G. Hu et al. (Eds.): CAINE 2024, CCIS 2242, pp. 115–126, 2025.
https://doi.org/10.1007/978-3-031-76273-4_9

potential to revolutionize cybersecurity education. By carefully crafting prompts, LLMs can provide personalized explanations, generate realistic attack scenarios, and create tailored learning experiences that meet users at their current skill level. Despite their popularity, successfully integrating LLMs into cybersecurity education presents unique challenges. These language models can be factually inaccurate, particularly in complex technical domains, potentially leading to the dissemination of incorrect or even harmful information [3]. To be reliable partners in learning, LLMs require a combination of well-structured, informative prompts and access to up-to-date threat intelligence. Moreover, the rapidly evolving cybersecurity landscape demands educational approaches that are equally dynamic and adaptable. Ethical considerations also come into play, as there is the potential for LLMs to be used for malicious purposes within the very domain they are meant to fortify. A prompt [17] is a set of instructions to produce customizable outputs for the LLM. However, poorly crafted prompts lead to unsatisfactory results. For instance, an overly broad prompt like "Tell us about cybersecurity" inundates users with excessive information, making it hard to extract relevant insights. This reinforces the need for focused and specific inquiries. Basic questions, such as "What is a phishing attack?", often yield superficial textbook definitions. Similarly, prompts like "Discuss the impact of cyberattacks on businesses" are a good start but fail to guide the LLM towards the most valuable insights. Overall, crafting effective cybersecurity prompts requires a nuanced understanding of both the domain knowledge and the LLM prompt layers. Otherwise, the users will find the experience frustrating or unproductive. To address this challenge, our research introduces a novel approach to prompt design: the prompt tree, a combination of *recursive fishbone diagrams*, as a systematic method for crafting informative prompts. This framework, grounded in trusted resources such as National Initiative for Cybersecurity Careers and Studies (NICCS) [9], the National Vulnerability Database (NVD) [20], the Common Vulnerabilities and Exposures (CVE) [19], the National Initiative for Cybersecurity Education Cybersecurity Workforce Framework (NICE) [10], and the Open Web Application Security Project (OWASP) [22], guides the creation of prompts specifically tailored for cybersecurity education, operations and methodologies. Our prompt design emphasizes adaptability, allowing users to explore diverse cybersecurity scenarios while maintaining a consistent learning structure. Our template "node" starts with attack that can be customized with specifics such as DDoS, SQL Injection, Phishing, etc. This is extended across various prompt dimensions, including vulnerabilities, skills, work roles, and certifications. We are generating a catalog of several such trees that the novice user can reference to for developing nuanced prompts. The user can use as-is or curtail such nodes based on the task objective such as designing incident management plan, hiring decisions etc. Our prompt tree allows users to explore diverse cybersecurity scenarios while maintaining a consistent learning structure. We have enhanced standard prompt engineering techniques such as Chain of Thought (CoT) to address diverse skill level and cybersecurity threat situations. The goal is to build more sophisticated, AI-enabled, and helpful products and services to democratize cybersecurity education and training. The main contributions of this paper are listed below:

- *Diverse Prompt Templates*: developed customizable prompt templates that facilitate the effective application of LLMs in cybersecurity education.
- *Catering Prompts for Diverse Security Skills Levels:* Our research introduces a framework that dynamically adapts prompts based on individual user skill levels. This ensures a personalized learning experience, providing tailored challenges to enhance Cybersecurity understanding.
- *Framework-Driven for Accuracy and Relevance:* Developed a system grounded in trusted frameworks like NICCS [9], NVD [20], CVE [19], NICE [10], and OWASP [22]. This ensures our prompts align with real-world cybersecurity concepts, standards, and in-demand skills.

This paper is organized into four main sections. Section 2 reviews the state-of-the-art in prompt engineering and its application to cybersecurity education. Section 3 details our methodology, including the development of the prompt tree framework and its integration with established cybersecurity resources. Section 4 presents the evaluation of our prompts and prompt sequences through diverse LLM testing and outlines our work on prompts. Finally, Sect. 5 concludes with a summary of our research contributions and a discussion of potential implications for the field of cybersecurity education.

2   Related Work

In this section, we present a summary of work on prompt engineering for LLM highlighting the best practices and gaps in the state-of-art with regards to democratizing cybersecurity education. Generative AI, especially LLMs [1,8,17], holds the promise of co-creation through human-machine interaction, facilitating new ways of communicating and interacting with these technologies [4,5]. Within this context, prompt engineering has emerged as a crucial discipline for maximizing the effectiveness of LLMs in knowledge transfer [24,25].

Research in prompt engineering has yielded valuable frameworks and guidelines. Mahmoud Bsharat *et al.* [6] provide a comprehensive set of principles for designing LLM prompts, while White *et al.* [25] propose a classification framework for prompt patterns, categorizing them into Input Semantics, Output Customization, Error Identification, Prompt Improvement, Interaction, and Context Control. This systematic approach provides a toolbox for refining prompts for specific LLM use cases. Liu *et al.* [18] examine techniques for bypassing safeguards in LLMs, highlighting potential risks and ethical considerations in educational settings.

Although these works touch on prompt design, a successful LLM-based learning tool must center the user experience, especially when empowering novice learners in complex fields such as cybersecurity. Ekin [13] offers an accessible guide to prompt engineering focused on ChatGPT, covering best practices and diverse use cases. Zamfrescu-Pereira *et al.* [26] investigate the differences between novice and expert prompt design, revealing challenges faced by non-technical users. Henrickson [15] explores the use of Natural Language Processing (NLP) techniques to enhance the clarity and fluency of prompts, which is valuable for making LLM-driven cybersecurity explanations more comprehensible.

The existing literature provides valuable insights into prompt engineering and the broader use of LLMs for education. However, a discernible gap exists regarding its application within the specialized and complex domain of cybersecurity education. Additionally, the emphasis often falls on expert refinement of prompts rather than on tools that empower novice users from the outset. *Our novel contribution addresses the unique challenges of teaching cybersecurity concepts to users with limited technical backgrounds using LLM.* By developing a framework designed to support novice users in crafting effective prompts, we aim to fill this crucial gap. This innovative approach unlocks the potential of LLMs for accessible cybersecurity training and education. ChatGPT said: ChatG

3   Prompt Tree Design for Cybersecurity Education

This section presents the detailed steps for our prompt tree design. We have leveraged well-established techniques such as the fishbone diagram and highly credible cybersecurity resources such as NVD, CVE, among others, to create the various nodes and structure of the tree.

3.1   Fishbone or Cause-and-Effect Diagram

The fishbone diagram, also known as the Ishikawa or cause-and-effect diagram, is a well-established problem-solving tool widely used across various domains. Its origins trace back to quality management in manufacturing, where it was employed to systematically identify the potential causes of product defects. It has been adapted for diverse applications, from identifying factors contributing to patient readmission in healthcare to uncovering root causes of low student engagement in education. Its flexible structure and visual representation make it an effective tool for brainstorming, problem-solving, and root cause analysis in complex systems.

In this research, we introduce a novel concept termed the *Prompt Tree*, directly inspired by the fishbone diagram's structure. We leveraged this framework to systematically identify and categorize the diverse factors that contribute to cybersecurity vulnerabilities and challenges. The "head" represents a specific cybersecurity problem or learning objective, while the main "bones" branch out to represent broad categories such as vulnerabilities, attack vectors, required skills, relevant work roles, and potential certifications. Each of these main categories is further dissected into smaller "sub-bones", representing more specific elements within each category.

To populate and refine our Prompt Tree, we employed a combination of data-driven and expert-informed approaches. We utilized topic modeling to analyze vast datasets of cybersecurity vulnerabilities, extracting recurring themes and key concepts. These insights guided the selection of relevant categories and sub-bones within our diagram. We also drew upon established frameworks like NICE, NICCS [9], OWASP [22], NVD [20], and CVE [19] to ensure our Prompt Tree aligned with real-world cybersecurity knowledge, skills, and career pathways. This iterative process of data analysis and expert input allowed us to create a comprehensive and adaptable framework for generating effective cybersecurity prompts tailored to users' diverse learning needs.

Table 1. PHASE I - Vulnerability Exploration and Analysis

Framework	How is it Used in Prompt Tree	Purpose of dimension
OWASP [22] (Open Web Application Security Project) Focuses on improving web application security.	*Attacks*: Informed by OWASP's attack methodologies, prompt development incorporates various threat vectors relevant to web application security (e.g., SQL injection, cross-site scripting).	Offers up-to-date insights on web application security risks and best practices.
NVD [20] (National Vulnerability Database) U.S. government repository of publicly disclosed cybersecurity vulnerabilities.	*Vulnerabilities*: Extracted from detailed datasets within NVD to populate the prompt tree. This ensures prompts address real-world cybersecurity weaknesses. For instance, to investigate various access-related vulnerabilities, one can navigate to the NVD. By utilizing the NVD search function and entering "access" as the keyword, the database will provide a list of vulnerabilities. Each entry will include details such as the Vulnerability ID (Vuln ID), a summary of the vulnerability, and its Common Vulnerability Scoring System (CVSS) severity rating.	Provides a comprehensive dataset of cybersecurity vulnerabilities with details and severity ratings.
CVE [19] (Common Vulnerabilities and Exposures) List of known vulnerabilities and exposures in information technology products.	*Vulnerabilities*: CVE serves as a resource alongside NVD for identifying and incorporating specific vulnerabilities into the prompt tree. For instance, to explore access-related vulnerabilities, navigate to the CVE database. By entering "access" in the search, the CVE database will display relevant vulnerabilities, including details such as CVE ID, a brief description, and references.	Establishes a standardized way to discuss and track vulnerabilities across cybersecurity tools.
ExploitDB [21] A database of known exploits for various vulnerabilities.	*ExploitDB* provides realistic examples of how vulnerabilities are exploited, enhancing the understanding of attack vectors for advanced users.	Provides real-world exploit examples for highly technical understanding (to be approached very carefully).

Our prompt tree methodology offers several distinct advantages. *Firstly*, it is adaptable and versatile to various security issues and practical needs. The multi-dimensional recursive structure allows for adaptation across various domains within and beyond cybersecurity. Users can tailor their approach by focusing on specific variables within a consistent, reliable framework. Any section of the tree can act as the starting point for exploration, and its branches can be dynamically customized to align with the specific learning needs or interests of the user.

Secondly, the methodology is scalable. The tree effectively accommodates problem spaces of varying sizes and complexity, remaining functional regardless of whether a user is exploring a few variables or many. *Thirdly*, it is user-friendly. The fishbone structure promotes intuitive understanding and requires minimal prior knowledge to use. This makes it accessible to a wide range of users, regardless of their technical background. *Lastly*, the prompt tree is multi-dimensional. Its value extends beyond cybersecurity, with its focus on root-cause analysis making it applicable to problem-solving in various fields.

3.2   Leveraging Trusted Frameworks

To ensure the accuracy and relevance of our prompt tree, we anchored it in several well-established cybersecurity frameworks. *NICE Framework:* mapped cybersecurity work roles and defined crucial knowledge, skills, and abilities (KSAs) essential for cybersecurity professionals [10]. *NICCS* illuminated workforce trends and identified high-demand skill areas in the cybersecurity landscape, providing valuable insights for our study [9]. *NVD & CVE:* These resources provided detailed breakdowns of common vulnerabilities, their potential impact, and associated exploits, offering essential data for our analysis [20] [19]. *OWASP* focused on web application vulnerabilities, informing prompts for secure coding practices, which are integral to cybersecurity education and training [22].

These resources, as summarized in Tables 1 and 2, provided the foundation for identifying real-world vulnerabilities and understanding the associated attack methodologies. Ultimately, it helps stakeholders such as CSOs, CISOs, and CEOs map the skills required for effective mitigation and defense. As shown in Table 1, we began by delving into attack methodologies detailed by OWASP

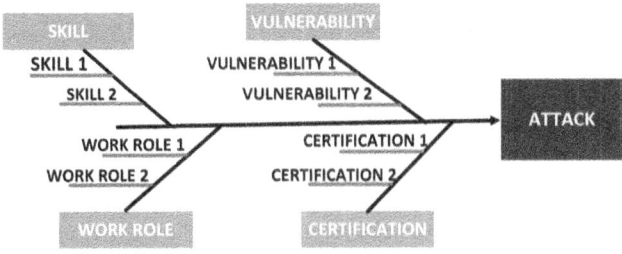

Fig. 1. Generic Prompt Tree

[22]. This informed our understanding of common threat vectors. Next, using the NVD [20], we extracted vulnerabilities associated with these specific attacks. We then cross-referenced these vulnerabilities with the CVE database [19], obtaining unique CVE identifiers and in-depth vulnerability descriptions. Finally, utilizing ExploitDB [21] and these CVE identifiers, we collected detailed exploit information, including targeted technologies and software.

In Table 2, we turned our attention to the human element of cybersecurity using the NICCS resource. This provided insights into workforce needs, career paths, and educational resources. We then cross-referenced the vulnerability data (from CVE and ExploitDB) with NICCS information, identifying specific work

Table 2. Phase II - Connecting Vulnerabilities, Work Roles, and Skills

Framework	How is it Used in Prompt Tree	Purpose of dimension
NICCS [9] (National Initiative for Cybersecurity Careers and Studies) Resources for cybersecurity education, training, and career development.	*Work Roles* (indirectly): NICCS resources likely informed the selection of relevant work roles identified within the NICE Framework (mentioned below).	Provides insights into cybersecurity education, training, and career development needs.
ExploitDB	Matching the vulnerabilities and details from ExploitDB in NICCS	This has helped us by finding out the relation between the vulnerabilities and workforce details.
NICE [10] (National Institute of Cybersecurity Education Framework) Knowledge, Skills, and Abilities (KSAs) needed for cybersecurity work.	*Work Roles*: Used to identify relevant cybersecurity job functions and their required skills. These roles informed the creation of prompts targeting specific areas of expertise (e.g., Penetration Tester, Security Analyst). *Skills*: Mapped to the prevention, detection, and response abilities outlined by NICE, ensuring prompts address critical cybersecurity competencies.	Provides a standardized taxonomy for cybersecurity workforce development and education.
CyberSeek [12] Resource that provides insights into current trends and needs in the cybersecurity job market.	*Work Roles*: Provides how different work roles need different certifications and what should be the skill level for that specific work role.	Shows real-world job market data for practical career guidance.

roles involved in managing these threats. The NICE Framework was then used to extract the knowledge, skills, and abilities (KSAs) essential for each work role. Finally, we integrated data from CyberSeek to understand how these work roles align with in-demand certifications and the required skill levels for success.

3.3   Prompt Tree Versions

Our generic prompt tree in Fig. 1 provides a versatile and scalable foundation for cybersecurity education. This flexibility allows us to create multiple variations tailored to specific user needs and goals. Figure 2 shows a tree focused on patching vulnerabilities, useful for mitigating known weaknesses. Alternatively, users can adapt the tree for hiring cybersecurity professionals or developing incident response plans by focusing on relevant work roles, skills, and certifications related to specific attacks. This customizable approach enables *personalized learning paths*, empowering users to navigate their cybersecurity training based on individual goals. They can use the prompt tree to address specific vulnerabilities, prepare for particular attacks, or build a career in cybersecurity.

Our research stands out by systematically integrating these previously isolated silos of fundamental security knowledge. By combining insights from diverse frameworks (OWASP [22], NVD [20], CVE [19], NICCS [9], NICE [10], CyberSeek [12]), we have developed a prompt tree that bridges the gap between technical vulnerabilities, real-world attack vectors, required skillsets, and career pathways. This holistic approach to prompt design for LLM-driven cybersecurity education represents a significant step towards making cybersecurity knowledge more accessible, relevant, and actionable for learners of all levels.

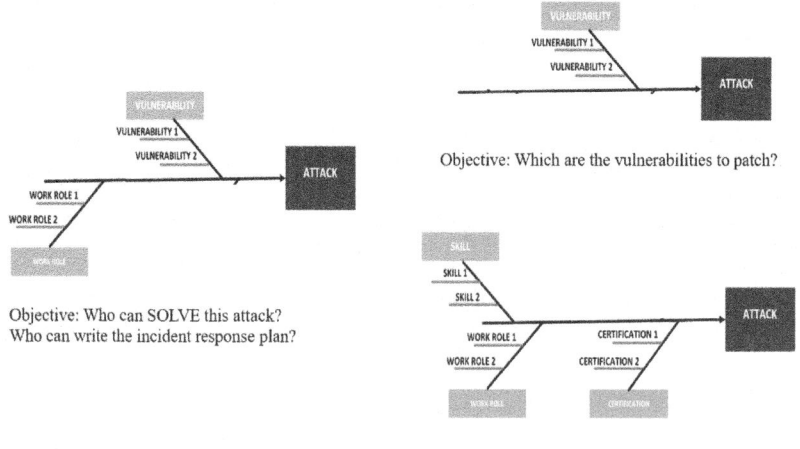

Fig. 2. The flexibility of the Prompt tree based on the user objective

4 Evaluation

In this evaluation, we present prompts derived from our fishbone diagram across five dimensions. We then designed and tested multi-tiered prompt sequences with leading LLMs, categorizing them as good, okay, or bad based on their efficacy and logical flow. This analysis, coupled with our user scenario (beginner, intermediate, expert), demonstrates the flexibility and effectiveness of our prompt tree in facilitating personalized and accessible cybersecurity education.

4.1 Prompt Sequence Evaluation

To optimize the learning experience, we employed a two-tiered prompt structure. *First-Level prompts* to provide foundational introductions to cybersecurity concepts and terminology. *Second-Level prompts* that delve into greater technical depth or focus on specific vulnerabilities and mitigation techniques. The user can extend into multiple levels inspired by our prompt tree from Sect. 3.3. Our fishbone diagram analysis guided the creation of over 50 prompts across 5 dimensions of the fishbone, which shows the flexibility of the prompt tree. We are sharing a brief subset for brevity in Table 3.

Table 3. Prompts for Attacks

Theme	Various Flavors for the same theme	Post Questions
Attacks	1. What premediate steps should we take for preparing for a potential [attack]?	1. If successful, which [attacks] would have the most severe impact?
	2. What are the immediate steps to take if we suspect an [attack] is in progress?	2. Who needs to be notified immediately, both internally and potentially externally?
		1. Which vulnerabilities may exist in our systems making them susceptible to [attack]?
		2. Which vulnerabilities pose the greatest risk if exploited in an [attack]?

To elevate our prompts, we then designed follow-up that can extend an initial user inquiry as shown in Table 4. We conducted a comprehensive analysis of *prompt sequences*, evaluating their efficacy and identifying patterns of successful interaction with leading LLMs. Crucially, we put these combinations to the test with three leading LLMs: ChatGPT, GEMINI, and Claude. This allowed us to observe how different LLMs respond and identify sequences that consistently yield high-quality results.

Prompt sequences were evaluated based on their logical flow, adaptability, and comprehensiveness within cybersecurity education. *Good sequences* exhibit a clear progression of inquiry, guiding the LLM through preparation, analysis,

Table 4. Sequence table

Type	Good sequence	Somewhat okay sequence	Bad sequence
Main flavor	What premediate steps should we take to prepare for a potential [attack]?		
Follow up-1	Which vulnerabilities exist in our systems, making them susceptible to specific attacks?	If successful, which [attacks] would have most severe impacts?	In a crisis, what [attacks] could someone with [skill] help to mitigate?
Follow up-2	Do we have the right [work roles] in place to develop and execute [attack] preparedness plan?	Who needs to be notified immediately, both internally and potentially externally?	What are the [vulnerabilities] that makes us more vulnerable for the [attack]?

and team readiness. An example of such a sequence begins with "What proactive steps can we take to prepare for a potential [attack]?", followed by "Which vulnerabilities in our systems make them susceptible to this attack?", and finally, "Do we have the right cybersecurity roles to develop and execute a preparedness plan?". The use of the '[attack]' placeholder allows for customization, making the sequence adaptable to specific threats. This well-designed structure not only enhances the quality and depth of insights from the LLM but also covers various security dimensions such as defense strategy, vulnerability assessment, and workforce considerations.

In contrast, *bad sequences* disrupt this logical flow, creating a disjointed learning experience for the LLM. For instance, a sequence starting with the same initial question, but followed by prompts like "If successful, which [attacks] would have the most severe impact?" and "Who needs to be notified immediately?", jumps between unrelated topics, neglecting crucial elements such as vulnerability assessment. This fragmented approach hinders effective learning and demonstrates the importance of careful prompt design. Analyzing both good and bad sequences provides valuable insights for refining the prompt creation process and ensuring a coherent and informative educational experience.

This treatment offer the user a systematic and transparent approach towards the prompt development process. We provide successful sequences along with the insights behind their design, fostering deeper understanding and encouraging customization within the learning experience. To illustrate the potential benefits of our system, consider a small business owner with limited security budget. Our prompts and the resulting interactions with the LLM could guide him/her in understanding phishing threats, recognizing suspicious emails, and improving his business's security posture. This highlights the real-world value of our approach in empowering non-technical users.

We evaluated our prompt tree with users across a range of cybersecurity experience levels. For beginners, prompts focus on foundational concepts and terminology, requiring no prior technical knowledge. Intermediate users are challenged with prompts that develop practical skills and address real-world scenarios. Finally, for seasoned experts, prompts encourage deeper analysis and exploration of cutting-edge threats and technologies.

5   Conclusion

The growing threat of cyberattacks necessitates innovative and inclusive approaches to cybersecurity education. This research addresses the critical knowledge gap faced by many individuals due to the technical nature of traditional cybersecurity resources. While prompt engineering offers vast potential, its specific application within cybersecurity education remains largely **underdeveloped**. This research has sought to address this gap by utilizing the power of Large Language Models (LLMs) to bridge this gap, empowering users to understand cybersecurity threats, develop practical skills, and proactively protect themselves in the digital landscape.

Our methodology focuses on developing a customizable prompt framework designed to translate complex cybersecurity concepts into actionable guidance for users with diverse backgrounds. The template-driven nature of our prompts offers significant adaptability. Users can personalize their learning experience by replacing placeholders like "[attack]" with specific threats (DDoS, SQLi, Phishing, etc.). This *flexibility*, applied across various prompt dimensions, empowers users to tailor their cybersecurity education based on individual needs and interests. It will also emphasize rigorous user testing through surveys and usability studies to gather feedback for continuous improvement of the prompt system. The future scope of this project envisions a comprehensive and adaptable framework of prompts that addresses the breadth of cybersecurity topics across various skill levels. This will be supported by robust infrastructure, ensuring the system is accessible and promotes intuitive interactions with LLMs.

Ultimately, this research aims to make cybersecurity education **accessible** to all. By democratizing this crucial knowledge, we empower both individuals and organizations to navigate digital threats and build a safer online world.

References

1. Artem, K., Sergiy, T.: Generative AI and prompt engineering in education. Mod. Eng. Innovative Technol. **1**(29–01), 117–121 (2023)
2. Asare, O., Nagappan, M., Asokan, N.: Is github's copilot as bad as humans at introducing vulnerabilities in code? (2024). arXiv preprint
3. Bommasani, R., Hudson, D.A., et al.: On the opportunities and risks of foundation models. ArXiv (2021). arXiv preprint
4. Bozkurt, A.: Generative AI, synthetic contents, open educational resources (OER), and open educational practices (OEP): A new front in the openness landscape. Open Praxis (2023)
5. Bozkurt, A.: Unleashing the potential of generative AI, conversational agents and chatbots in educational praxis: A systematic review and bibliometric analysis of GenAI in education. Open Praxis (2023)
6. Bsharat, S.M., Myrzakhan, A., Shen, Z.: Principled instructions are all you need for questioning LLaMA-1/2, GPT-3.5/4 (2024). arXiv preprint
7. Carleton, A., Klein, M., et al.: Robert. Architecting the Future of Software Engineering: A National Agenda for Software Engineering, Research & Development. Carnegie Mellon University, Software Engineering Institute's Digital Library (2021). Accessed 10 Jun 2024

8. Chen, B., Zhang, Z., Langrené, N., Zhu, S.: Unleashing the potential of prompt engineering in large language models: a comprehensive review (2023). arXiv preprint
9. Cybersecurity and Infrastructure Security Agency. National initiative for cybersecurity careers and studies. https://niccs.cisa.gov/ (2024). Accessed 10 Jun 2024
10. Cybersecurity and Infrastructure Security Agency. National initiative for cybersecurity careers and studies: Nice framework. https://niccs.cisa.gov/workforce-development/nice-framework (2024). Accessed 10 Jun 2024
11. Cybersecurity Ventures. Hackerpocalypse: A cybercrime report. https://cybersecurityventures.com/hackerpocalypse-cybercrime-report-2016/ (2016). Accessed 27 May 2024
12. CyberSeek. Cybersecurity supply/demand heat map (2024). https://www.cyberseek.org/. Accessed 10 Jun 2024
13. Ekin, S.: Prompt engineering for ChatGPT: A quick guide to techniques, tips, and best practices (2023)
14. GitHub. GitHub Copilot: Your AI Pair Programmer (2023). https://github.com/features/copilot. Accessed 10 Jun 2024
15. Henrickson, L., Meroño-Peñuela, A.: Prompting meaning: a hermeneutic approach to optimising prompt engineering with ChatGPT. AI & SOCIETY, pp. 1–16 (2023)
16. Krochmalski, J.: IntelliJ IDEA Essentials. Packt Publishing Ltd (2014)
17. Liu, P., Yuan, W., et al.: Pre-train, prompt, and predict: A systematic survey of prompting methods in natural language processing (2021). arXiv preprint
18. Liu, Y., Deng, G., Xu, Z., et al.: Jailbreaking ChatGPT via prompt engineering: An empirical study (2024). arXiv preprint
19. MITRE Corporation. Common vulnerabilities and exposures: Search CVE list (2024). https://cve.mitre.org/cve/search_cve_list.html. Accessed 10 Jun 2024
20. National Institute of Standards and Technology. National vulnerability database: Vulnerability search (2024). https://nvd.nist.gov/vuln/search. Accessed 10 Jun 2024
21. Offensive Security. Exploit database (2024). https://www.exploit-db.com/. Accessed 10 Jun 2024
22. Open Web Application Security Project. Owasp top ten project (2024). https://owasp.org/www-project-top-ten/. Accessed 10 Jun 2024
23. Pearce, H., Ahmad, B., et al.: Asleep at the keyboard? Assessing the security of github copilot's code contributions (2021). arXiv preprint
24. Velásquez-Henao, J.D., Franco-Cardona, C.J., Cadavid-Higuita, L.: Prompt engineering: a methodology for optimizing interactions with AI-language models in the field of engineering. DYNA **90**(230), 9-17 (2023)
25. White, J., Fu, Q., et al.: A prompt pattern catalog to enhance prompt engineering with ChatGPT (2023). arXiv preprint
26. Zamfirescu-Pereira, J.D., Wong, R.Y., Hartmann, B., Yang, Q.: Why Johnny can't prompt: How non-AI experts try (and fail) to design LLM prompts. CHI '23, New York, NY, USA (2023). Association for Computing Machinery

Parallel Computing and Algorithms

Bayesian Parameter Inference in Stochastic Biochemical Models Using Moment Approximations

Kannon Hossain$^{(\boxtimes)}$ (iD) and Roger B. Sidje$^{(\boxtimes)}$ (iD)

Department of Mathematics, The University of Alabama, Tuscaloosa,
AL 35487, USA
khossain@crimson.ua.edu, roger.b.sidje@ua.edu

Abstract. The chemical master equation (CME) is a mathematical tool utilized to model the stochasticity of the complex biochemical reaction networks. As the direct solution of the CME is notoriously expensive, moment-based approximations are computationally attractive in terms of time and memory to compute the statistics of the CME. The Bayesian method with delayed rejection adaptive Metropolis (DRAM) sampler provides an accurate probabilistic framework for parameter inference, reducing the need for costly computations of likelihood functions. In this study, we derive a system of ordinary differential equations (ODEs) using zero-moment approximations and employ the DRAM sampler algorithm to find the approximate posterior distributions of the model parameters. The quality of the method is evaluated through two systems biology case studies, with the aim of enabling efficient data-driven inferences for more complex biochemical models.

Keywords: Bayesian inference · Markov chain Monte Carlo (MCMC) · moment approximations · chemical master equation

1 Introduction

The behavior of the biochemical reaction networks at the cellular level is widely recognized to be stochastic, and the chemical master equation (CME) [1] is commonly used to describe the dynamics of complex biochemical reaction networks. Because of the 'curse of dimensionality', the size of CME can be extremely large or theoretically infinite, thus often not directly solvable. In the past few years, several approximation methods [2–6] have been implemented for solving the CME. Moment approximations [7,8] have gained much attention in recent years for approximating the statistical moments of the CME that proved to be numerically cheap for capturing the randomness of the stochastic models. The analysis of stochastic biochemical models necessitates precise measurement of model parameters, which are often only approximately known. Therefore, in statistics and machine learning, it is of common interest to address the inverse problem: using noisy observations of the time course data of a system to quantify the uncertainty of the model parameters. Studies have demonstrated the success

© The Author(s), under exclusive license to Springer Nature Switzerland AG 2025
G. Hu et al. (Eds.): CAINE 2024, CCIS 2242, pp. 129–143, 2025.
https://doi.org/10.1007/978-3-031-76273-4_10

of inferences based on ordinary least squares (OLS) [9] and maximum likelihood estimation (MLE) [10–13] for both deterministic and stochastic systems. The Bayesian approach utilized with various Markov chain Monte Carlo (MCMC) samplers [14–17] has been studied extensively for approximating the posterior distribution of the model parameters. Because of the iterative nature, MCMC-based algorithms such as the classical Metropolis-Hastings (MH) [18,19] sampler suffer from slow convergence and have difficulties finding the best proposal to fit the target density. The delayed rejection adaptive Metropolis (DRAM) [20] template, a combination of delayed rejection (DR) [21,22] and adaptive Metropolis (AM) [23] ensures the best fit of the proposal distribution to its target density by faster convergence. The AM algorithm uses cumulated information at the start of the simulation and calculates the covariance of the proposal density using all previous states, which reduces the multiple function evaluations and ensures an effective search process. On the other hand, the DR algorithm reduces the number of rejected candidates by proposing a new candidate instead of staying in the same state for a long time, which further reduces the correlation between states of the Markov chain.

Our method of estimation is twofold: first, we use zero-moment approximations to find the ODEs for the first moment and second central moment for each molecular species, which can later be solved numerically by MATLAB's *ode15s* [24]. Secondly, we employ the DRAM sampler to find the approximate posterior distribution of the model parameters embedded in the ODE systems. The test results from two case studies, namely, the birth-death and the dimerization processes, were shown to be efficient in inferring the posterior distribution of the model parameters.

We organize the remainder of the paper as follows: Sect. 2 provides methodological backgrounds on the CME, moment-based approximations, Bayesian inference, Metropolis-Hastings with adaptive Metropolis sampler, and the delayed rejection Metropolis-Hastings sampler with zero-moment approximations. Sections 3 describe the case studies with results and discussions. Lastly, concluding remarks on Sect. 4.

2 Background

2.1 Chemical Master Equation

Consider a chemical reaction system consisting of N molecular species that interact through M reactions of the form

$$R_k : a_{1k}S_1 + \cdots + a_{Nk}S_N \xrightarrow{c_k} b_{1k}S_1 + \cdots + b_{Nk}S_N \tag{1}$$

where a_{ik} and b_{ik} are coefficients denoting the number of S_i molecules, with c_k being the reaction rate constant for $1 \leq k \leq M$ and $1 \leq i \leq N$. We denote $\boldsymbol{x}(t) = (x_1, \ldots, x_N)^T$ as the state of the system at time t. The propensity function $\alpha_k(\boldsymbol{x}(t))$ of reaction R_k at the current state $\boldsymbol{x}(t)$ is defined so that the probability of such a reaction occurring during the infinitesimal time interval $[t, t + dt)$ is

$\alpha_k(\boldsymbol{x}(t))dt$. When reaction R_k happens, the state vector is updated with the stoichiometric vector $\boldsymbol{\nu}_k$, representing the change in species numbers.

Denote $P(\boldsymbol{x}, t) = \text{Prob}\{\boldsymbol{x}(t) = \boldsymbol{x}\}$, the probability that the system is at state $\boldsymbol{x}$ at time t. As given in Eq. [1], a characterization of CME is

$$\frac{dP(\boldsymbol{x}, t)}{dt} = \sum_{k=1}^{M} \alpha_k(\boldsymbol{x} - \boldsymbol{\nu}_k) P(\boldsymbol{x} - \boldsymbol{\nu}_k, t) - \sum_{k=1}^{M} \alpha_k(\boldsymbol{x}) P(\boldsymbol{x}, t) \qquad (2)$$

Let $\boldsymbol{X}$ be the set of all possible states, if we order these states as $\boldsymbol{X} = \{\boldsymbol{x}_1, \ldots, \boldsymbol{x}_n\}$, where $\boldsymbol{x}_i = (x_{1i}, \ldots, x_{Ni})^T$ and n is the total number of states, then Eq. (2) defines a set of ODEs

$$\dot{\boldsymbol{p}}(t) = \boldsymbol{A} \cdot \boldsymbol{p}(t) \qquad t \in [0, t_f] \qquad (3)$$

where $\boldsymbol{A}$ is the transition rate matrix. From Eq. (3) the probability vector at the end point t_f is

$$\boldsymbol{p}(t_f) = \exp(t_f \boldsymbol{A}) \, \boldsymbol{p}(0) \qquad (4)$$

Here, for the purpose of parameter estimation, we will leverage moment approximations to simulate the dynamics of the system by generating ODEs instead of finding the solution of the entire probability distribution of the CME.

2.2   Moment-Based Approximations

We denote $\mathbb{E}[x_i] = \mu_i$ as the first moment of the i-th species. To obtain it, we multiply Eq. (2) by x_i and summing over all reachable states $\boldsymbol{x} = (x_1, \ldots, x_N)^T$

$$\sum_{\boldsymbol{x} \in X} x_i \frac{dP(\boldsymbol{x}, t)}{dt} = \sum_{\boldsymbol{x} \in X} \left(\sum_{k=1}^{M} x_i \alpha_k(\boldsymbol{x} - \boldsymbol{\nu}_k) P(\boldsymbol{x} - \boldsymbol{\nu}_k, t) \right.$$

$$\left. -x_i P(\boldsymbol{x}, t) \sum_{k=1}^{M} \alpha_k(\boldsymbol{x}) \right) \qquad (5)$$

For the second central moment, multiply Eq. (2) by $(x_i - \mu_i)(x_j - \mu_j)$ and summing over all reachable states $\boldsymbol{x} = (x_1, \ldots, x_N)^T$

$$\sum_{\boldsymbol{x} \in X} (x_i - \mu_i)(x_j - \mu_j) \frac{dP(\boldsymbol{x}, t)}{dt} =$$

$$\sum_{\boldsymbol{x} \in X} \left(\sum_{k=1}^{M} (x_i - \mu_i)(x_j - \mu_j) \alpha_k(\boldsymbol{x} - \boldsymbol{\nu}_k) P(\boldsymbol{x} - \boldsymbol{\nu}_k, t) \right.$$

$$\left. - (x_i - \mu_i)(x_j - \mu_j) P(\boldsymbol{x}, t) \sum_{k=1}^{M} \alpha_k(\boldsymbol{x}) \right) \qquad (6)$$

Following the derivation given in [25], where we apply a transformation $\boldsymbol{x} - \boldsymbol{\nu}_k \rightarrow \boldsymbol{x}$ using the fact that $\mathbb{E}\left[\alpha_k(\boldsymbol{x})\right] = \sum_{\boldsymbol{x} \in X} \alpha_k(\boldsymbol{x}) P(\boldsymbol{x}, t)$ with $\alpha_k(\boldsymbol{x})$ representing the kth reaction propensity at state $\boldsymbol{x}$, therefor Eqs. (5) and (6) can be written as

$$\frac{d\mathbb{E}\left[x_i\right]}{dt} = \sum_{k=1}^{M} \nu_{k,i} \mathbb{E}\left[\alpha_k(\boldsymbol{x})\right] \tag{7}$$

$$\frac{d\mathbb{E}\left[(x_i - \mu_i)(x_j - \mu_j)\right]}{dt} = \sum_{k=1}^{M} \Big(\nu_{k,i} \mathbb{E}[(x_j - \mu_j)\alpha_k(\boldsymbol{x})]$$

$$+ \nu_{k,j} \mathbb{E}\left[(x_i - \mu_i)\alpha_k(\boldsymbol{x})\right] + \nu_{k,i}\nu_{k,j} \mathbb{E}\left[\alpha_k(\boldsymbol{x})\right] \Big) \tag{8}$$

The moment Eqs. (7) and (8) generate an infinite set of coupled non-linear ODEs. A moment closure truncates the infinite ODEs to a finite one by eliminating the dependence of lower-order m^{th} moments on higher-order $(m + 1)^{th}$ moments that are usually intractable and cannot be solved analytically or numerically. The zero-moment approximation is a basic moment approximation method where the Taylor series is truncated at a specific order, as if all subsequent higher-order moments were set to be zero. Our focus will be on the zero-moment approximation or zero closure, where the highest moment order is considered to be two.

If we denote $\mathbb{E}\left[x_i\right] = \mu_i$ and $\mathbb{E}\left[(x_i - \mu_i)(x_j - \mu_j)\right] = \sigma_{ij}$ and applying multivariate Taylor series expansion of $\mathbb{E}\left[\alpha_k(\boldsymbol{x})\right]$ around the mean $\boldsymbol{\mu} = (\mu_1, \ldots, \mu_N)^T = (\mathbb{E}\left[x_1\right], \ldots, \mathbb{E}\left[x_N\right])^T$, by setting the third central moment equal to zero, one can simplify Eqs. (7) and (8) as

$$\frac{d\mu_i}{dt} = \sum_k \nu_{k,i} \left(a_k(\boldsymbol{\mu}) + \frac{1}{2} \sum_{l,m} \frac{\partial^2 a_k(\boldsymbol{\mu})}{\partial x_l \partial x_m} \sigma_{lm} \right) \tag{9}$$

$$\frac{d\sigma_{ij}}{dt} = \sum_k \left[\nu_{k,i} \left(\sum_l \frac{\partial a_k(\boldsymbol{\mu})}{\partial x_l} \sigma_{jl} \right) + \nu_{k,j} \left(\sum_l \frac{\partial a_k(\boldsymbol{\mu})}{\partial x_l} \sigma_{il} \right) \right.$$

$$\left. + \nu_{k,i}\nu_{k,j} \left(a_k(\boldsymbol{\mu}) + \frac{1}{2} \sum_{l,m} \frac{\partial^2 a_k(\boldsymbol{\mu})}{\partial x_l \partial x_m} \sigma_{lm} \right) \right] \tag{10}$$

where μ_i is the first moment and σ_{ij} is the second central moment. Later on, Eqs. (9) and (10) can be used to derive the ODEs for each species of our test models. A detailed derivation of moment closure approximations can be found in these articles [7, 8, 25].

2.3   Bayesian Inference

The need for parameter estimation arises from the fact that, as Eqs. (9) and (10) are utilized to derive the explicit ODEs for each species (for example, see

Eqs. (18), (19), (20) and (21)) with some unknown reaction rate parameters $\boldsymbol{\theta} = (c_1, \ldots, c_M)^T$ embedded in the system need to be estimated. Unlike point-based estimations mentioned earlier (e.g., OLS, MLE) that seek to obtain point estimates of $\boldsymbol{\theta}$, the goal of the Bayesian method is to approximate the posterior probability distribution of $\boldsymbol{\theta}$. Given the prior distribution $\pi(\boldsymbol{\theta})$, Bayes' theorem can be expressed as a proportionality:

$$\pi(\boldsymbol{\theta} \mid \mathcal{D}) \propto \pi(\mathcal{D} \mid \boldsymbol{\theta}) \pi(\boldsymbol{\theta}) \tag{11}$$

where $\pi(\boldsymbol{\theta} \mid \mathcal{D})$ is the posterior distribution, and $\pi(\mathcal{D} \mid \boldsymbol{\theta})$ is the likelihood of $\boldsymbol{\theta}$ given the data $\mathcal{D}$.

2.4 Metropolis-Hastings and the Adaptive Metropolis Sampler

The MH algorithm generates a Markov chain, dividing the sampling process into two steps: the proposal step and the probabilistic acceptance-rejection step. It generates a chain on the parameter space $\boldsymbol{\theta}$, and the steps are as follows:

(I). Choose initial parameter $\boldsymbol{\theta}_0$ and proposal density $q(\cdot \mid \boldsymbol{\theta}_0)$.
(II). Using current value of the chain $\boldsymbol{\theta}_i$, propose a new candidate $\boldsymbol{\theta}^*$ using proposal density $q(\cdot \mid \boldsymbol{\theta}_i)$.
(III). Compute the acceptance ratio

$$\alpha(\boldsymbol{\theta}_i, \boldsymbol{\theta}^*) = \min\left(1, \frac{\pi(\boldsymbol{\theta}^*) \, q(\boldsymbol{\theta}_i \mid \boldsymbol{\theta}^*)}{\pi(\boldsymbol{\theta}_i) \, q(\boldsymbol{\theta}^* \mid \boldsymbol{\theta}_i)}\right) \tag{12}$$

(IV). Generate a uniform random number $u \in [0, 1]$
  – If $u \leq \alpha(\boldsymbol{\theta}_i, \boldsymbol{\theta}^*)$, then accept the candidate by setting $\boldsymbol{\theta}_{i+1} = \boldsymbol{\theta}^*$
  – If $u > \alpha(\boldsymbol{\theta}_i, \boldsymbol{\theta}^*)$, then reject the candidate by setting $\boldsymbol{\theta}_{i+1} = \boldsymbol{\theta}_i$
(V). Go to step (II). until enough values have been sampled.

There are numerous ways [26] to select the proposal density q, and here we only consider the symmetric case with the Gaussian proposal distribution (Gaussian likelihood), i.e., if the current state of the chain is at $\boldsymbol{\theta}$, we have

$$q(\boldsymbol{\theta}^* \mid \boldsymbol{\theta}) \propto \exp\left(-\frac{1}{2}(\boldsymbol{\theta}^* - \boldsymbol{\theta})^T \boldsymbol{\Sigma}^{-1}(\boldsymbol{\theta}^* - \boldsymbol{\theta})\right) \tag{13}$$

where $\boldsymbol{\Sigma}$ is a positive definite matrix that determines the covariance of the proposal distribution. With this choice of a symmetric proposal distribution, the MH algorithm reduces to the original Metropolis algorithm [18]. The selection of an effective proposal distribution for the Metropolis algorithm is crucial for achieving efficient results through simulations, requiring the appropriate choice of Σ. The adaptive Metropolis (AM) algorithm improves the MH algorithm by updating the proposal Σ every step, eliminating the need for users 'tuning' the distribution. As the covariance of the proposal distribution of the AM adaptation depends on the history of the chain, we assume that after an initial non-adaptation period, the Gaussian proposal is centered at the current position of the Markov chain $\boldsymbol{\theta}_i$. In particular, let $\boldsymbol{\theta}_1, \ldots, \boldsymbol{\theta}_i$ be the samples accepted so far. To initiate the adaptation procedure, we choose an arbitrary strictly positive definite initial Gaussian proposal density covariance, $\boldsymbol{\Sigma}_0$, based on our prior knowledge. The AM updates the proposal covariance using the following formula:

$$\boldsymbol{\Sigma} = \boldsymbol{\Sigma}_i := \begin{cases} \boldsymbol{\Sigma}_0, & i \leq n_0 \\ s_d \operatorname{Cov}(\boldsymbol{\theta}_1, \ldots, \boldsymbol{\theta}_i) + s_d \varepsilon \boldsymbol{I}_d, & i > n_0 \end{cases} \tag{14}$$

Here, the function Cov returns the sample covariances. The constant s_d depends only on the dimension d of the state space with the assigned value $(2.4)^2/d$, where n_0 is the number of initial steps without proposal adaptations, which can be freely chosen. $\boldsymbol{I}_d$ denotes the d-dimensional identity matrix, and we choose $\varepsilon = 10^{-6}$ very small to ensure that $\boldsymbol{\Sigma}_i$ will not become singular.

2.5 Delayed Rejection Metropolis-Hastings Sampler

The main idea behind the delayed rejection (DR) algorithm [21] over MH is to reduce the correlation between the states of the Markov chain. Whenever a candidate is rejected, instead of taking the current state of a Markov chain as its new state, it proposes a new candidate. This process can be iterated to reduce the number of rejected candidate states, which will improve the standard MH algorithm. The DR steps are as follows:

(I). Let the current position of a sampled chain be at $\boldsymbol{\theta}_i = \boldsymbol{\theta}$. Suppose a candidate move $\boldsymbol{\theta}_1^*$ is generated from a proposal density $q(\cdot \mid \boldsymbol{\theta})$ and accepted with the usual MH algorithm

$$\alpha_1(\boldsymbol{\theta}, \boldsymbol{\theta}_1^*) = \min\left(1, \frac{\pi(\boldsymbol{\theta}_1^*) \, q_1(\boldsymbol{\theta} \mid \boldsymbol{\theta}_1^*)}{\pi(\boldsymbol{\theta}) \, q_1(\boldsymbol{\theta}_1^* \mid \boldsymbol{\theta})}\right) \tag{15}$$

(II). If rejected, instead of staying at the same position, $\boldsymbol{\theta}_{i+1} = \boldsymbol{\theta}$, a second stage move, $\boldsymbol{\theta}_2^*$, is proposed. The second stage proposal is allowed to depend on both the current position and what we have just proposed and rejected: $q(\cdot \mid \boldsymbol{\theta}, \boldsymbol{\theta}_1^*)$ and we calculate the accepted probability with

$$\begin{aligned} \alpha_2(\boldsymbol{\theta}, \boldsymbol{\theta}_1^*, \boldsymbol{\theta}_2^*) = \min\Bigg(1, &\frac{\pi(\boldsymbol{\theta}_2^*) \, q_1(\boldsymbol{\theta}_1^* \mid \boldsymbol{\theta}_2^*)}{\pi(\boldsymbol{\theta}) \, q_1(\boldsymbol{\theta}_1^* \mid \boldsymbol{\theta})} \times \\ &\frac{q_2(\boldsymbol{\theta} \mid \boldsymbol{\theta}_1^*, \boldsymbol{\theta}_2^*) \, [1 - \alpha_1(\boldsymbol{\theta}_2^*, \boldsymbol{\theta}_1^*)]}{q_2(\boldsymbol{\theta}_2^* \mid \boldsymbol{\theta}, \boldsymbol{\theta}_1^*) \, [1 - \alpha_1(\boldsymbol{\theta}, \boldsymbol{\theta}_1^*)]}\Bigg) \end{aligned} \tag{16}$$

This process of delaying rejection can be iterated to try sampling from further proposals in case of rejection by the present one; thus, a proposal can be simulated up to s-stages.

2.6 Delayed Rejection Adaptive Metropolis Sampler

The efficiency of the DR process can be improved by adaptation when good proposals are unavailable. As DR utilizes a predetermined number of proposals that are employed at different stages, the success greatly depends on the assumption that at least one of the proposals is successfully calibrated. On the other hand, AM provides a methodical solution when the adaptation process begins slowly. The concept of AM strategies involves learning from the information gathered during the run of the chain to optimize proposals efficiently. Combining AM with DR can be done in various ways [20], but a direct method using an s-stages DR setup is preferred here:

(I). In the first stage of DR, the proposal is adapted just as in AM: the covariance Σ_i^1 is computed from the points of the sampled chain, no matter at which stage these points have been accepted in the sample path.

(II). The covariance Σ_i^r of the proposal for the r-th stage ($r = 2, \ldots, s$) is always computed simply as a scaled version of the proposal of the first stage, $\Sigma_i^r = \gamma_r \Sigma_i^1$.

The scale factor γ_r can be selected without any restrictions. According to the simulation presented in [22], with an appropriate choice of γ_r, one can reduce the asymptotic variance of the resulting estimators more effectively by keeping greater variance at earlier stages and then reduce the variance by rejecting.

2.7   Delayed Rejection Adaptive Metropolis with Zero-Moment Approximations

The optimal performance of the MCMC method depends on how well the proposal distribution fits the target distribution; therefore, choosing the initial covariance for the proposal distribution is important for the convergence of the chain. For example, if the initial guess for the proposal density is far from the correct one, i.e., the variance of the proposal is too large or its covariance is almost singular, no proposal will be accepted, thus the adaptation process does not get started. Instead of manually tuning the proposal density, our method starts first by finding a point estimate for both of the parameters using least squares (LS) estimation. In order to initiate the MCMC chain and prevent the long burn-in, the best-fit parameters returned from the fitting scheme are used as the initial parameter values for the MCMC run. The residual error returned from the fit is used as the initial model variance. The initial proposal covariance (jump) is based on the approximated covariance matrix returned by the fitting scheme scaled with $(2.4)^2/d$, which gives a suitable covariance matrix for generating new parameter values. The number of accepted runs can be increased by taking appropriate DR steps with at least 2 and by updating the proposal distribution after a certain iteration. Equations (9) and (10) are used to derive the closed-form ODEs of first moment and second central moment for each molecular species that can be solved numerically by MATLAB's built-in solver *ode15s*. It is important to note that some people refer to the first moment as "mean" and the second central moment as "variance." Readers should differentiate these terms from the fact that model parameters can also have posterior means. Noisy observed time course data can be generated by Gillespie's SSA [6] algorithm, which we called SSA sample mean and SSA sample variance. A standard OLS fitting scheme would be:

$$\boldsymbol{\theta}_{\text{LS}} = \arg\min_{\boldsymbol{\theta}} \left[(\hat{\boldsymbol{\mu}} - \boldsymbol{\mu}(\boldsymbol{\theta}))^2 + (\hat{\boldsymbol{\sigma}} - \boldsymbol{\sigma}(\boldsymbol{\theta}))^2 \right] \tag{17}$$

where $\boldsymbol{\mu}$ is the first moment, $\hat{\boldsymbol{\mu}}$ is the SSA sample mean, $\boldsymbol{\sigma}$ is the second central moment, and $\hat{\boldsymbol{\sigma}}$ is the SSA sample variance of each species. The objective function described by Eq. (17) can be minimized by a derivative-free optimizer *fminsearch*, available in MATLAB. Both the proposal densities and the likelihood function are considered to be Gaussian in the DRAM sampler. The prior distributions for all the model parameters chosen to be uniform, i.e., $\boldsymbol{\theta} \sim U(a, b)$, where a and b represent the minimum and maximum values of the parameters, which vary depending on the models.

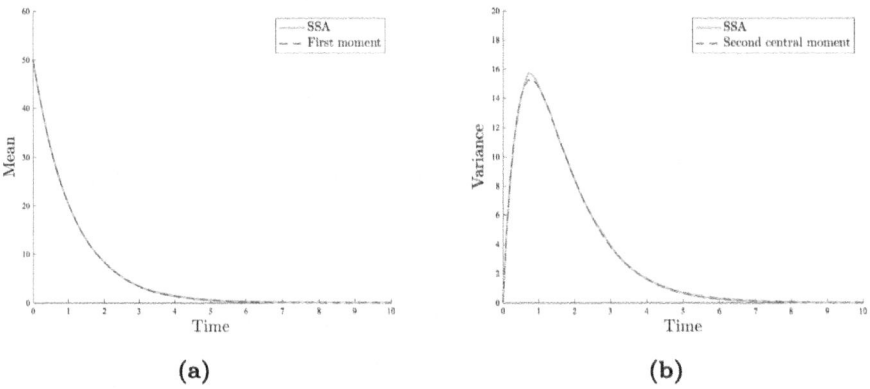

Fig. 1. Simulation of the birth-death process: (a) mean of protein; (b) variance of the protein.

3   Case Studies

3.1   Birth-Death Process

Table 1. Birth-death process

	Reaction	Propensity	State Change Vectors
R_1:	$X \xrightarrow{c_1} 2X$	$\alpha_1 = c_1 x$	$\nu_1 = 1$
R_2:	$X \xrightarrow{c_2} \varnothing$	$\alpha_2 = c_2 x$	$\nu_2 = -1$

Consider the following model given in Table 1 for the production and degradation of a protein species X with a molecule x, where in reaction R_1 protein is produced at a constant rate c_1 and in reaction R_2 protein is degraded at a constant rate c_2. By using Eqs. (9) and (10) one can find following the sets of ODE

$$\frac{d\mu_1(t)}{dt} = (c_1 - c_2)\,\mu_1(t), \mu_1(0) = x_0 \tag{18}$$

$$\frac{d\sigma_{11}(t)}{dt} = 2\,(c_1 - c_2)\,\sigma_{11}(t) + (c_1 + c_2)\,\mu_1(t), \sigma_{11}(0) = 0 \tag{19}$$

where μ_1 is the first moment and σ_{11} is the second central moment of protein. We sampled 100 data points by running 10^4 SSA realizations and added Gaussian noise $\mathcal{N}(0, 0.5^2)$ by choosing the true parameter values: $\boldsymbol{\theta}_{true} = (c_1, c_2) = (0.1, 1.0)$ [27] and known initial conditions $(\mu_1(0), \sigma_{11}(0)) = (50, 0)$ for the time $t = 10s$. The solutions of Eqs. (18) and (19), together with the data set, can be used to formulate the objective function given by Eq. (17). With the initial guess chosen as $\boldsymbol{c_0} = (2.0, 5.0)^T$, the minimizer *fminsearch* provides the point estimate of each parameter (shown as vertical dashed lines in Fig. 2), which is later used as the initial parameter values for the MCMC run as noted before. Of course, one can chose other biologically plausible initial guesses for c_0 to run the optimizer from a specific interval. For a low-dimensional problem, the choice of $n_0 = 100$ ensured a smooth start to the non-adaptation period. The number of MCMC runs is chosen to be 5000, where the DR sampler executes a 2-stage proposal with an adaptation interval of 100. The prior distribution for both of the parameters c_1 and c_2 is considered to be uniformly distributed with $c_1 \sim U(0, 2)$ and $c_2 \sim U(0, 5)$.

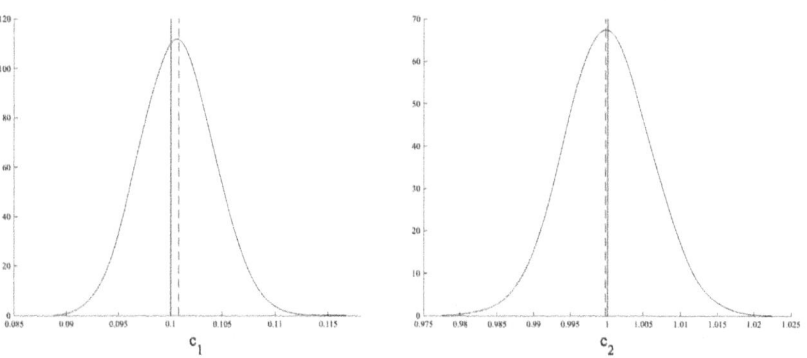

Fig. 2. Posterior distribution (smooth black curve) of the parameters c_1 and c_2 of the birth-death process, where the vertical solid lines show the true values. The vertical dashed lines show the point estimates returned by the LS method using Eq. (17).

Table 2. Posterior mean and standard deviation of the parameters of the birth-death process.

Parameter	True Values	Mean	Standard deviation
c_1	0.1	0.10059	3.44×10^{-3}
c_2	1.0	0.9995	5.51×10^{-3}

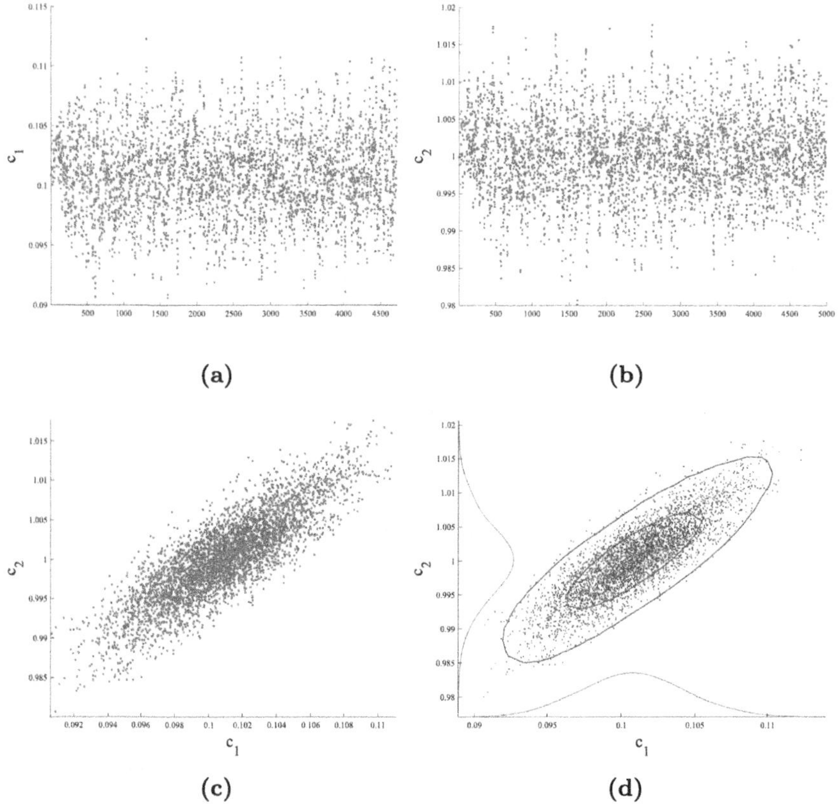

(a)

(b)

(c)

(d)

Fig. 3. (a), (b) showing one-dimensional MCMC chains of the parameters c_1 and c_2 of the birth-death process; (c) two-dimensional MCMC chains of c_1 and c_2; (d) two-dimensional posterior distributions of c_1 and c_2 where the dots give the points of the MCMC chain from which the distribution contour lines (for the inner 50% and outer 95% regions of the distribution) are calculated. The distributions along the axes represent one-dimensional marginal densities (with c_1 along the x-axis and c_2 along the y-axis).

Results and Discussions. The initial proposal variance can be too small or too large, causing difficulties in AM adaptation and slow convergence. The DRAM sampler offers a significant advantage in DR implementation, as it utilizes the covariance of the fit proposal in the first stage, scaled by $(2.4)^2/2 = 2.88$

for further improvement. The acceptance rates of the chain were found to be 85.66%. Figure 2 depicts the posterior distribution of the parameters c_1 and c_2 compared to the true values. The posterior mean and standard deviation for each parameter are shown in Table 2, which further indicates the successful implementation of the DRAM sampler as the mean of each parameter is close to true values with very low standard deviations. The mixing of the chain is crucial for improved MCMC sampling. The DRAM sampler ensures that the chain is well mixed using the proper DR steps and AM adaptation. We can clearly see that Fig. 3(a) and 3(b) depict the well mixing of the chains of parameters c_1 and c_2. Figure 3(c) and 3(d) show the two-dimensional chains for both of the parameters, which are well identified and positively correlated.

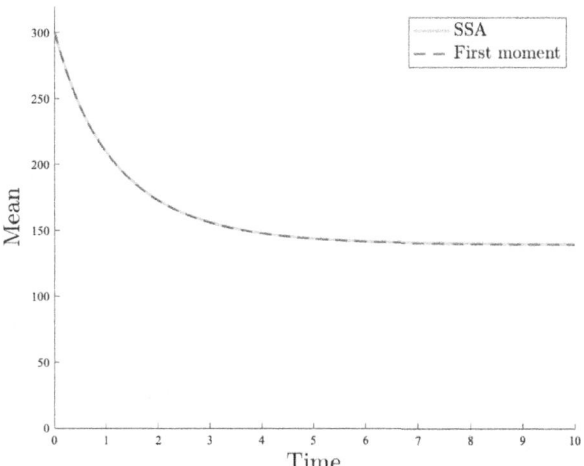

Fig. 4. Simulation of the dimerization process: mean of the monomer.

3.2 Dimerization Process

Table 3. Dimerization process

	Reaction	Propensity	State Change Vectors
R_1:	$2X \xrightarrow{c_1} Y$	$\alpha_1 = c_1 \frac{x(x-1)}{2}$	$\nu_1 = (-2, 1)$
R_2:	$Y \xrightarrow{c_2} 2X$	$\alpha_2 = c_2 \frac{(x_0 - x)}{2}$	$\nu_2 = (2, -1)$

Consider the reversible dimerization process given in Table 3, where two monomers X can fuse into a single dimer Y and the dimer can split apart into two monomers. With Eqs. (9) and (10), we can derive the following sets of ODE:

$$\frac{d\mu_1(t)}{dt} = c_1\mu_1(t)\left(1 - \mu_1(t)\right) + c_2\left(x_0 - \mu_1(t)\right) - c_1\sigma_{11}(t), \mu_1(0) = x_0 \quad (20)$$

$$\frac{d\sigma_{11}(t)}{dt} = -2c_1\left(2\mu_1(t) + 2\right)\sigma_{11}(t) - 2c_2\sigma_{11}(t) + 2c_1\mu_1(t)\left(\mu_1(t) - 1\right)$$
$$+ 2c_2\left(\mathbb{X}_0 - \mu_1(t)\right), \sigma_{11}(0) = 0 \quad (21)$$

where μ_1 is the first moment and σ_{11} is the second central moment of the continuous state variable $\mathbb{X}(t)$ of the monomers. Note that the second-order moment truncation leaves the covariance between the monomers and dimers to zero. We will show that the ODEs for the monomers are sufficient to estimate the posterior distributions of the model parameters. We sampled 100 data points by running 10^4 SSA realizations and added Gaussian noise $\mathcal{N}(0, 0.5^2)$ by choosing the true parameter values: $\boldsymbol{\theta}_{true} = (c_1, c_2) = (0.00166, 0.2)$ [28] and known initial conditions $(\mu_1(0), \sigma_{11}(0)) = (301, 0)$ for the time $t = 10s$. The numerical solutions of Eqs. (20) and (21), along with the data set, are employed to construct the objective function in Eq. (17), where MATLAB's *fminsearch* optimizer is implemented as a minimizer. We run the optimizer with the initial guess $\boldsymbol{c_0} = (0.01, 1.0)^T$ to find the point estimate of the parameters (shown as vertical dashed lines in Fig. 5). We later used the optimizer's optimal value as the initial parameter value for the MCMC run. The non-adaptation point is set to be $n_0 = 100$, and we allowed 5000 total MCMC runs, where the DR sampler executed a 2-stage proposal with an adaptation interval of 100. The prior distribution for both parameters c_1 and c_2 is considered to be uniformly distributed, with $c_1 \sim U(0, 0.5)$ and $c_2 \sim U(0, 2)$. All of the parameters are log-transformed, i.e., log_{10} is used in MATLAB for calculation and graphing purposes.

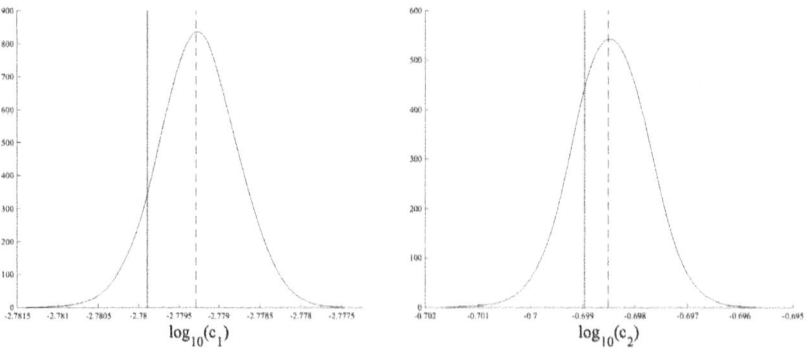

Fig. 5. Posterior distribution (smooth black curve) of the parameters c_1 and c_2 of the dimerization process, where the vertical solid lines show the true values. The vertical dashed lines show the point estimates returned by the LS method using Eq. (17).

Table 4. Posterior mean and standard deviation of the parameters of the dimerization process.

Parameter	True Values	Mean	Standard deviation
$\log_{10}(c_1)$	-2.779	-2.7793	4.58×10^{-4}
$\log_{10}(c_2)$	-0.6995	-0.69852	6.85×10^{-4}

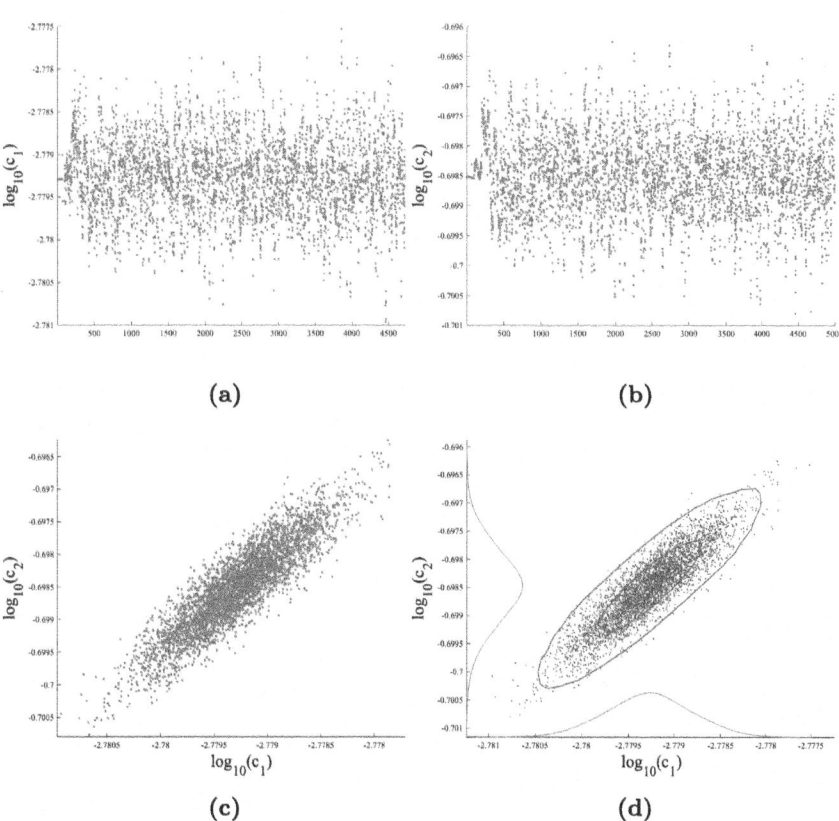

Fig. 6. (a), (b) showing one-dimensional MCMC chains of the parameters c_1 and c_2 of the dimerization process; (c) two-dimensional MCMC chains of c_1 and c_2; (d) two-dimensional posterior distributions of c_1 and c_2 where the dots give the points of the MCMC chain from which the distribution contour lines (for the inner 50% and outer 95% regions of the distribution) are calculated. The distributions along the axes represent one-dimensional marginal densities (with c_1 along the x-axis and c_2 along the y-axis).

Results and Discussions. Figure 5 depicts the posterior distribution of the parameters c_1 and c_2 compared to the true parameters. The posterior mean and standard deviation for each parameter are shown in Table 4, indicating the successful implementation of the DRAM sampler as the mean of each parameter is close to true values with very low standard deviations. Since the model has only two parameters ($d = 2$), we scaled the returned covariance matrix by $2.4^2/2$ = 2.88. The acceptance rates of the chains was found to be 84.55%. The well-mixing of the chain for both parameters c_1 and c_2 are shown in Fig. 6(a) and 6(b). Two-dimensional chains for both of the parameters are graphed in Fig. 6(c) and 6(d), which demonstrates that the chains are well identified and positively correlated.

4   Conclusion

In this study, we investigated the use of the MCMC-based Bayesian parameter inference method to find the posterior distribution of the parameters in stochastic biochemical models. Our findings demonstrate the successful combination of two modifications to the standard MH sampler: AM, which adapts the proposal distribution based on the chain's past history, and DR, which enhances the efficiency of the MCMC estimator. The multivariate Gaussian proposals were used due to their ease of fit, but other proposals can also be utilized within the proposed framework. We have also shown that the zero-moment approximations, which are computationally much faster and cheaper than expensive solutions of the CME, work well with the DRAM sampler. Since the integration of moment approximations is highly efficient and often fast compared to other methods, we intend to explore more complex biochemical systems with higher-order moment conditions combined with different MH samplers to find the posterior distributions of the model parameters.

References

1. Gillespie, D.T.: A rigorous derivation of the chemical master equation. Phys. A Stat. Mech. Appl. **188**(1-3), 404–425 (1992)
2. Munsky, B., Khammash, M.: The finite state projection algorithm for the solution of the chemical master equation. J. Chem. Phys. **124**(4), 044104 (2006)
3. Kazeev, V., Khammash, M., Nip, M., Schwab, C.: Direct solution of the chemical master equation using quantized tensor trains. PLoS Comput. Biol. **10**(3), e1003359 (2014)
4. Vo, H.D., Sidje, R.B.: An adaptive solution to the chemical master equation using tensors. J. Chem. Phys. **147**(4), 044102 (2017)
5. Dinh, T., Sidje, R.B.: An adaptive solution to the chemical master equation using quantized tensor trains with sliding windows. Phys. Biol. **17**(6), 065014 (2020)
6. Gillespie, D.T.: Exact stochastic simulation of coupled chemical reactions. J. Phys. Chem. **81**(25), 2340–2361 (1977)
7. Engblom, S.: Computing the moments of high dimensional solutions of the master equation. Appl. Math. Comput. **180**(2), 498–515 (2006)
8. Lee, C.H.: A moment closure method for stochastic chemical reaction networks with general kinetics. MATCH Commun. Math. Comput. Chem. **70**, 785–800 (2013)
9. Hossain, K., Sidje, R.B.: Parameter estimation in biochemical models using moment approximations. In: 2023 International Conference on Computational Science and Computational Intelligence (CSCI), pp. 551–557. IEEE (2023)

10. Hossain, K., Sidje, R.B.: Case-studies of parameter estimation in the stochastic reaction-diffusion master equation. In: Proceedings of the 16th International Conference on Bioinformatics and Computational Biology, vol. 101, pp. 103–112 (2024)
11. Simpson, M.J., Browning, A.P., Warne, D.J., Maclaren, O.J., Baker, R.E.: Parameter identifiability and model selection for sigmoid population growth models. J. Theor. Biol. **535**, 110998 (2022)
12. Dinh, K.N., Sidje, R.B.: An application of the Krylov-FSP-SSA method to parameter fitting with maximum likelihood. Phys. Biol. **14**(6), 065001 (2017)
13. Hossain, K., Sidje, R.B.: Parameter estimation in biochemical models using marginal probabilities. In: Southwest Data Science Conference, pp. 197–211. Springer (2023)
14. Zeng, T., Spence, J.P., Mostafavi, H., Pritchard, J.K.: Bayesian estimation of gene constraint from an evolutionary model with gene features. Nat. Genet. **56**, 1–12 (2024)
15. Särkkä, S., Svensson, L.: Bayesian filtering and smoothing, volume 17. Cambridge University Press (2023)
16. Vo, H.D., Fox, Z., Baetica, A., Munsky, B.: Bayesian estimation for stochastic gene expression using multifidelity models. J. Phys. Chem. B **123**(10), 2217–2234 (2019)
17. Catanach, T.A., Vo, H.D., Munsky, B.: Bayesian inference of stochastic reaction networks using multifidelity sequential tempered markov chain monte carlo. Int. J. Uncertainty Quantification **10**(6), 515–542 (2020)
18. Metropolis, N., Rosenbluth, A.W., Rosenbluth, M.N., Teller, A.H., Teller, E.: Equation of state calculations by fast computing machines. J. Chem. Phys. **21**(6), 1087–1092 (1953)
19. Hastings, W.K.: Monte carlo sampling methods using markov chains and their applications (1970)
20. Haario, H., Laine, M., Mira, A., Saksman, E.: Dram: efficient adaptive MCMC. Stat. Comput. **16**, 339–354 (2006)
21. Mira, A., et al.: On metropolis-hastings algorithms with delayed rejection. Metron **59**(3–4), 231–241 (2001)
22. Green, P.J., Mira, A.: Delayed rejection in reversible jump metropolis–hastings. Biometrika **88**(4), 1035–1053 (2001)
23. Haario, H., Saksman, E., Tamminen, J.: An adaptive metropolis algorithm. Bernoulli, pp. 223–242 (2001)
24. MATLAB. version 7.10.0 (R2022a). The MathWorks Inc., Natick, Massachusetts (2022)
25. Lee, C.H., Kim, K.-H., Kim, P.: A moment closure method for stochastic reaction networks. J. Chem. Phys. **130**(13), 134107 (2009)
26. Roberts, G.O., Rosenthal, J.S.: General state space markov chains and MCMC algorithms (2004)
27. Wilkinson, D.J.: Stochastic modelling for quantitative description of heterogeneous biological systems. Nat. Rev. Genet. **10**(2), 122–133 (2009)
28. Ale, A., Kirk, P., Stumpf, M.P.H.: A general moment expansion method for stochastic kinetic models. J. Chem. Phys. **138**(17), 174101 (2013)

Introducing Multidimensional Parallel Decoder to Reduce Cost of Implementation and Latency

Nagi Mekhiel$^{(\boxtimes)}$

Department of Electrical, Computer and Biomedical Engineering, Toronto
Metropolitan University, Toronto, Canada
nmekhiel@ee.torontomu.ca
http://www.ee.torontomu.ca/nmekhiel

Abstract. The latency and complexity of decoders are critical to performance of devices such as memory and buffers. The number of locations to be accessed in an address could be in billions. We propose a multidimensional parallel decoder that divides the address space into multiple of smaller dimensions each is decoded separately in parallel. The decoding of a much smaller address is simpler compared to the decoding of the whole address space. A combinational circuit combines the outputs of the smaller numbers of decoded outputs that correspond to the different dimensions to obtain the full decoded address space. We also purpose a time multiplexed decoder that divides the address to multiple dimensions in time and uses the multidimensional parallel decoder to obtain the decoded outputs. The results of the multidimensional parallel decoder show reduction in the cost of implementation and latency by multiple folds compared to the conventional decoder.

Keywords: memory systems · latency of memory access · complexity of implementation · performance of storage systems · decoder design

1 Introduction

The processor uses an address to select a single location from a wide range of address space to access data from the selected location. For example, a thirty bit address produces an address range of one billion locations. Only one of these locations is selected by the processor. The speed of the decoder to select a single location is critical for fast retrieval of data and transferring it to the processor. Not only the delay of the decoder is important but also its complexity. The size of implementation affects the cost, the size, the reliability and power consumption.

The inventor of [2] proposed fast cyclic decoder using a clock generator to generate a clock and uses flip flops to capture the address. It uses another set of static flip flops with combinational logic to generate the word lines. This approach is suitable for FIFO/LIFO Data Buffers and increases the complexity of the implementation and latency due to clocking.

G. Hu et al. (Eds.): CAINE 2024, CCIS 2242, pp. 144–152, 2025.
https://doi.org/10.1007/978-3-031-76273-4_11

The inventor of [3] proposed a row decoder circuit that divides the global row address into multiple stages which increases the delay of decoding a global address.

The inventor of [4] proposed reducing decoder delay in memory devices. This method uses complicated design and adds latches to generate an effective address. This applies only to DRAM technology and is used to hide some of the delay in activating DRAM rows.

This work proposes a new method reduces the complexity and latency of the decoder which is used in many devices as different types of memory and input/output. The reduction of complexity improves power consumption and reduces the cost of implementation also reduces latency therefore, improves the access time of these devices.

Therefore, it is useful to optimize both the delay and the size of the decoder as it is used in many devices as memory, buffers and any storage devices.

2   Motivations

Present decoders combine all signals of the full address space to produce one active signal for the selected address. The circuit that evaluates the full address space is therefore complicated and it increases exponentially with the increase in the address range. It also takes longer time to produce the selected decoded active signal. Dividing the full address into multiple smaller addresses that could be evaluated at the same time in parallel reduces the complexity and latency.

3   The Concept of Multidimensional Parallel Decoder

Figure 1 shows a block diagram of the multidimensional parallel decoder. Processor P address is divided into number of smaller addresses or dimensions. The address is then decoded separately in parallel using simpler decoder. For example, a 24 bits address space has 16 million locations, and could be divided into four dimensions each has 6 bits address. Each decoder will then select only 64 locations therefore the complexity of the main decoder is reduced by 256,000 times. A combinational circuit combines the decoded outputs of the limited number of smaller parallel decoders to generate the outputs to the full address space. The complexity of the combinational circuit depends on the number of dimensions and not on the number of bits of the full address as in conventional decoder.

4   The Implementation of Multidimensional Decoder

Figure 2 shows the implementation of the multidimensional parallel decoder. The address space is assumed to be 24 bits A23..A0 range. The multidimensional parallel decoder divides the address space into four different dimensions each consists of 6 bits. The first dimension has A5..A0, the second from A11..A6, the

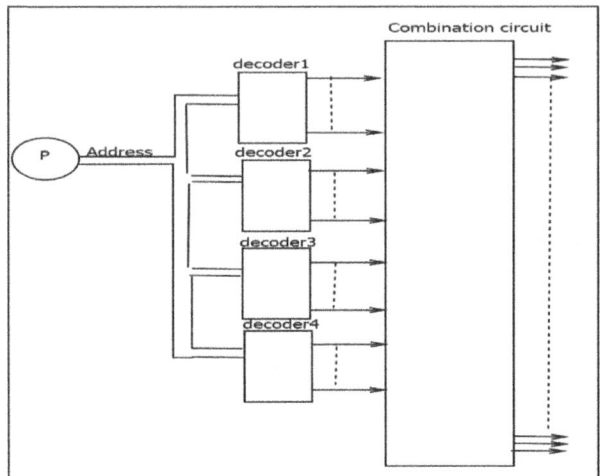

Fig. 1. The Concept of the Multidimensional Parallel Decoder

third from A17...A12 and the fourth from A23..A18. The number of dimensions could vary depending on the desired optimization in design and implementation. Each dimension address is applied to a much smaller decoder producing 64 outputs Q0..Q63. All of these decoders work in parallel, therefore the delay is only dependent on decoding 6 inputs compared to the delay of decoding 24 inputs for full address space. Therefore, the delay of each dimensional decoder is multiple times faster than the delay of decoding the full address as the delay of a logic gate is proportional to the number of inputs.

A combinational circuit combines the outputs from each dimension and produces the decoded output of the full address space. The number of inputs to produce one decoded output is only equal to the number of dimensions. The gate that combines each four inputs from each dimension is a simple AND gate and is simpler than the gates used to decode 24 inputs. All outputs for the full address space from the AND gates are also decoded in parallel. The delay of the full address is dependent on the delay of a one dimensional decoder of 6 inputs plus the delay of the combinational circuit of 4 inputs. The total number of inputs is 10 gate delay compared to 24 inputs gate delay for the conventional decoders.

5   Evaluating The Complexity of Decoders

5.1   Complexity of the Conventional Decoder

Figure 3 shows the schematic of a 24 bits conventional decoder implemented using 4 inputs AND gates. The number of decoded outputs is 16 million and each output needs to decode 24 bits of inputs. The first stage needs 6 gates, then second stage needs two AND gates followed by third stage of 1 AND gate

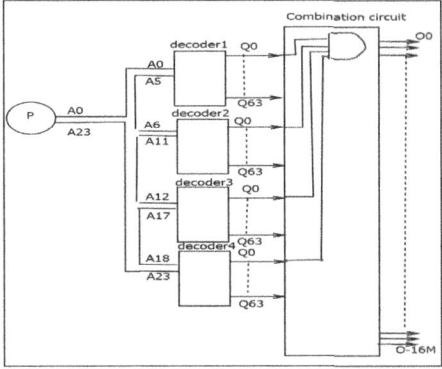

Fig. 2. Implementation of multidimensional parallel decoder

to generate each output. Therefore number of gates needed = (6+2+1) x16 million= 144 million of 4 inputs AND gates.

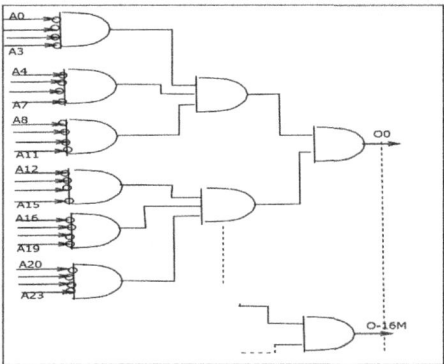

Fig. 3. Schematic of a conventional decoder

Figure 4. shows a block diagram of a two level conventional decoder. The first level uses a 12 bits decoder followed by the second level that uses 4096 of 12 bits decoders to generate 16 million outputs. Each of the 12 bits decoder uses 3 of the 4 inputs AND gate in the first stage followed by one AND gate to generate an output for the 4096 outputs. The total number of gates in each of conventional 12 bits decoder = (3+1)x4096= 16,000 AND gates. The two level conventional decoder uses 1 +4096 =4097 decoders each uses 16,000 gates. Total number of gates = 4KX16K= 64 million gates of four inputs AND gates.

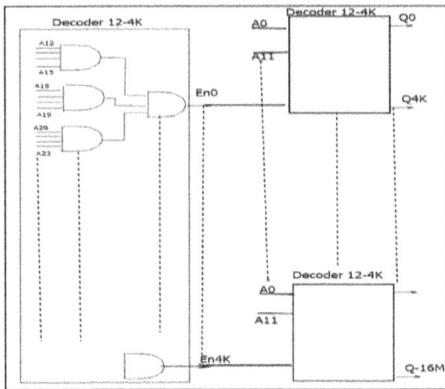

Fig. 4. Block Diagram of a two level conventional decoder

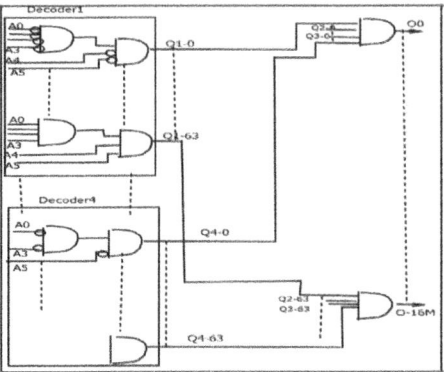

Fig. 5. Implementation of the multidimensional parallel decoder

5.2   The Complexity of Multidimensional Parallel Decoder

Figure 5 shows the implementation of the multidimensional parallel decoder using four inputs AND gates. The 24 bits address is divided into 4 dimensions each has 6 bits and uses a decode of 64 outputs. The inputs of the first dimension assigned A5..A0 and its outputs assigned Q63..Q0. The second dimension decoder inputs assigned to A11..A6 and outputs Q63..Q0, third has its inputs assigned to A17..A12 and outputs Q63..Q0, the fourth has its inputs assigned to A23..A18 and outputs Q63..Q0. The combinational circuit combines each four outputs from the corresponding four dimensions parallel decoders to generate a single output from O-0 to O-16million. O0 is generated from combining Q0 of all 4 dimensions using AND gate. O-16M is generated from combining Q63 outputs of the four dimensions decoders using AND gate. The number of gates in each single dimension decoder = 2 × 64. The total number of gates in 4 dimensions decoders = 4 × 2x64= 512 AND gates. Combination circuit uses 1 gate per each output, total $= 1 \times 16$ million$= 16$ million gates. Total number of gates in the multidimensional decoder $= 512 + 16$ million. The reduction in complexity of our multidimensional parallel decoder compared to the conventional decoder is 9 times and 4 times compared to the two level conventional decoder.

6 The Time Multiplexed Multidimensional Parallel Decoder

The time multiplexed parallel decoder divides the address space to multiple dimensions that are then used by the multidimensional parallel decoder with multi-phase clocking. The processor sends one dimension of the address each time synchronized by a multi-phase clocking. Multiple latches store the address of each dimension at the proper time. The advantage of this time multiplexed decoder is to have a much narrower bus interface from a processor to a memory. The draw back is increasing the delay of this decoder due to the time needed to latch the full address. However, the use of multidimensional parallel decoder can compensate for the increase in the latching delay as shown in next section.

Figure 6 shows a block diagram of a time multiplexed multidimensional parallel decoder. The processor address is divided into 4 equal addresses that each is sent at specific phase of clocking to be latched by the corresponding latch. Latch0 stores the first portion of the address A7..A0 for a 32 bits processor address, then Latch1 stores address A15..A8 at second phase of clocking, followed by latching A23..A16 at the third phase then A31..A24 at the fourth phase. Following the latching of the full address is the multidimensional parallel decoder that decodes the address of each latch in parallel as mentioned above then the combination circuit that combines the decoded outputs to produce the full decoded address space. The complexity of this time multiplexed multidimensional parallel decoder is the same complexity and cost of the multidimensional parallel decoder plus three latches. The cost of latches is very minimum compared to the cost of the decoder.

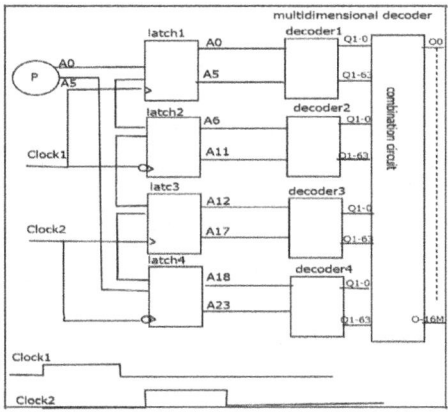

Fig. 6. Block diagram of a time multiplexed multidimensional parallel decoder

7 Analysis of the Complexity and Latency for the Multidimensional Parallel Decoder

We assume that the number of address lines to be decoded equal N bits. Assuming that the number of dimensions used to divide the address lines equal K.

7.1 Complexity Reduction of the Conventional Decoder

The conventional decoder have a number of outputs that is equal to 2^N and each output must decode the full address of N bits using AND function [1].

The complexity of the conventional decoder Cconv could be obtained as:-

$$Cconv = 2^N \times N \tag{1}$$

The multidimensional parallel decoder has K of parallel decoders each decodes N/K address bits for the first stage.

The second stage of multidimensional parallel decoder consists of 2^N outputs and each output combines K number of signals using AND function.

The complexity of multidimensional parallel decoder Cmulti could be obtained as:-

$$Cmulti = 2^{N/K} \times N/K + 2^N \times K \tag{2}$$

The gain in the reduction of complexity compared to conventional decoder is therefore given by the following equation:-

$$Cgain = Cconv/Cmulti \tag{3}$$

7.2 Latency Reduction of the Conventional Decoder

We assume that the delay of AND gates used to implement decoders is proportional to its number of inputs or Fan- in [1].

The latency of conventional decoder Delayconv could be obtained as:-

$$Delay.conv = N \tag{4}$$

The latency of multidimensional parallel decoder consists of adding the delays of the first stage which is a decoder of N/K inputs plus the delay of combination circuit that has K inputs.

Therefore the latency of multidimensional parallel decoder: -

$$Delay.multi = N/K + K \tag{5}$$

The gain in the reduction of latency is therefore given by the following equation:-

$$Delay - gain = N/(N/K + K) \tag{6}$$

7.3 The Results and Discussions

Figure 7 shows the gain in complexity reduction for using multidimensional parallel decoder compared to the conventional decoder for dimensions K=2, K=4 and K=8 as N changes from 8 to 32 bits. The best gain is achieved with number of dimensions =2. The reason for the decrease in gain when the number of dimensions increases is that the complexity of the combinational circuit increases linearly as K increases which is given by $2^N \times K$ and the gain becomes limited to N/K .

Increasing the number of address lines N for a decoder increases the gain in complexity reduction linearly making the multidimensional decoder suitable for decoders with large address space.

It should be noted that even for large number of dimensions when K=8, with N=32 bit, the gain in complexity reduction is four times that of conventional decoder.

Figure 8 shows the gain in the reduction of decoder delay for using multidimensional parallel decoder compared to the conventional decoder for K=2, K=4, and K=8 and N changes from 8 to 32 bits. The results show that the best gain in delay reduction is achieved with large number of dimensions equal 8. The gain starts to decrease as the number of address lines N increases because N/K from the multidimensional first stage decoder increases linearly with N.

The gain in delay reduction is limited to less than 2 for two dimensions, but the gain for eight dimensions is about 2.5.

The choice of the number of dimensions is dependent on the application. If it requires reduction on complexity, then it should use less number of dimensions and if it requires less delay then it uses large number of dimensions.

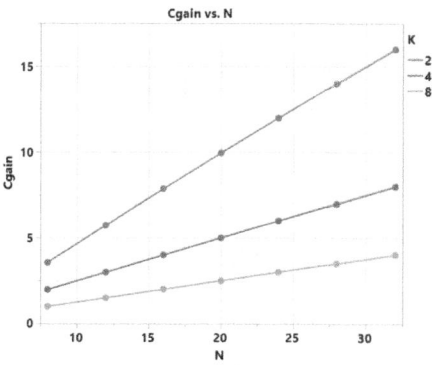

Fig. 7. Reduction in Complexity Using Multidimensional Parallel Decoder

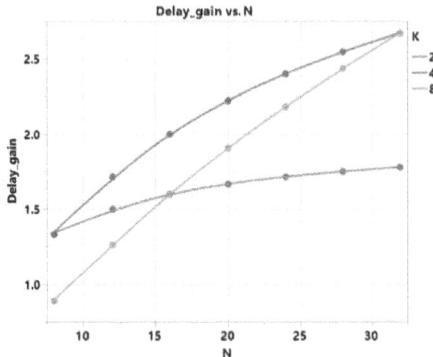

Fig. 8. Reduction in Delay Using Multidimensional Parallel Decoder

8   Conclusions

The multidimensional parallel decoder reduces the complexity and delay of the conventional decoder by multiple times. Time multiplexing of the address combined with multidimensional parallel decoder reduces the bus interface width and the cost of decoder compared to the conventional decoder.

The method could be used for future simpler and faster memory devices that are essential for all different types of computers and also it could be applied to optimize performance of some searching algorithms.

References

1. Brown, S., Vranesic, Z.: Fundamentals of Digital Logic with VHDL Design, Third Edition, McGraw-Hill (2009)
2. Nagendra Chandrakar Advanced Micro Devices, Inc. "Fast cyclic decoder circuit for FIFO/LIFO data buffer", US8238187B2
3. Im Cheol Ha, SK Hynix Inc "Decoder circuit used in a flash memory device" , US6870769 B1
4. Bae, Y.-C.: Samsung Electronics Co Ltd "Reduced delay address decoders and decoding methods for integrated circuit memory devices" US 6181635 B1

Introducing Novel Parallel Computing Using Orbital Data

Nagi Mekhiel$^{(\boxtimes)}$

Department of Electrical, Computer and Biomedical Engineering, Toronto
Metropolitan University, Toronto, Canada
nmekhiel@ee.torontomu.ca
http://www.ee.torontomu.ca/nmekhiel

Abstract. We propose a novel parallel computing that allows processors
to access data in predictable time without the need to access it from
different locations in memory using addresses. It uses orbital data that
is mapped to time and is made available to multiple processors at the
same time in multiple different orbits and at a specific predictable time
in each orbit. This allows processors in different orbits to share the same
data, eliminating the problem of sharing data at the same time among
multiple processors.

It provides processors with the ability to hide the waiting time when
accessing shared data by overlapping it with useful work on another data
while allowing other processors to work on the shared data in another
orbit.

The performance of this novel method shows significant improvements
in scalability compared to that of conventional parallel computing.

Keywords: Data accesses · Computer models · Parallel Computers ·
Memory systems · Scalability of parallel Computing

1 Introduction

Conventional Computers' performance is limited by the memory bottleneck and
cannot continue to improve with the advancements in technology.

The current parallel computing uses multiple processors working in parallel to execute an application, suffers from the overhead of synchronization and
communications between processors when sharing data. The multiprocessor with
shared memory requires synchronization for accessing shared data which limits
the performance gain and scalability [1,2,6,7]. Multiprocessors with distributed
memory uses messages to communicate between processors. The cost of these
messages limits the performance of this system [2,6,7].

It therefore, not a good idea to force many processors to work together at the
same time on the same data that maps to the same locations to reduce the time
of work and it is better if we allow the same data to be available for multiple
processors at the same time in multiple places or orbits. This enables processors
to escape the conflict of having to share the data that maps to the same space
at the same time.

© The Author(s), under exclusive license to Springer Nature Switzerland AG 2025
G. Hu et al. (Eds.): CAINE 2024, CCIS 2242, pp. 153–163, 2025.
https://doi.org/10.1007/978-3-031-76273-4_12

2   The Motivations for Using Orbital Data in Parallel Computing

Many scientific and engineering applications use data as input for their models to provide results. Machine learning uses data analysis to automate model building. The model can learn from data, identify patterns, and make decisions with minimal or no human intervention. The iterative aspect of machine learning is important because as models are exposed to new data, they can be able to independently adapt. Models learn from previous computations to produce reliable, more accurate decisions and results, therefore easy and fast access to big data is important.

Artificial intelligence applications depend on data to make fast and accurate decisions. The accuracy and effectiveness of AI depends on accessing big data multiple times.

2.1   The Requirements of Data for Efficient Computing

- Data must be able to expand therefore should not be stored in fixed locations and should not be mapped to space. Adding more data should be easy and should be done without interrupting the system if data is mapped to time using other orbits.
- Data should be available in a periodic cyclic fashion such that the model can access the same data multiple times to give a chance for the processor to analyze it and make accurate prediction as in ML algorithm.
- The same data should be applied to multiple different algorithms at the same time working in parallel for different models and analyze the data then give several results and outcomes if data is replicated in different orbits.
- It should limit the overhead of transferring data from one processor to another if the same data is delivered at a specific predictable time in multiple orbits.

3   The Concept of Orbital Data

We invented a Time-Based Access Memory "TAM" that maps data to time not to space as in the conventional memory. TAM could be used for easy and fast access to data [3,4].

Rather than having one processor to access one location (space) in memory at a time, we allow all the data to be available to all processors in a predictable time. The processor waits for TAM to provide its data each at a specific time [3,4].

The contents of memory are available to all processors using a shared bus or orbit. Every location in memory is guaranteed to be delivered to the bus so that processors have all the data available after waiting for a time that does not exceed the time of transferring the memory section out in the orbit. The period of the orbital cycle depends on the number of locations and bus speed. There are multiple orbits with different number of memory locations and different cycle times that are known to each processor.

The orbital memory makes its content spin continuously around multiple orbits, so that any processor can access the required data at specific periodic time. This memory uses time as an address for the processor to access data and is considered as a Time Addressable Memory "TAM" [3,4].
It has the following Features: -

– It supplies all contents of the memory to all processors all the time, regardless of if a processor needs it or not using a shared bus (orbit).
– It is accessed sequentially in time and does not need an address to access it as in the conventional memory.
– The access of each location is known ahead of time, so there is no waiting time for decoding and accessing data.
– It could use DRAM or SRAM or optical memory technology with very simple organization without decoders.

Figure 1. shows a block diagram of this concept. The contents of a memory section or bank spin around all the time at a fast speed and is delivered to any processor with each element at a specific time.

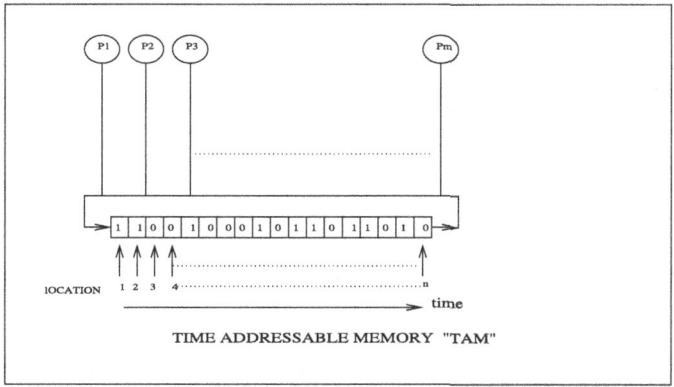

Fig. 1. The Concept of Orbital Data

3.1   Multi Level Orbital Data

Multi-Level Orbital Data provides access to different sections of memory with different cycle times. A section has several number of memory locations accessed in a serial or sequential order (mapped in a linear time). Cycle time is the time it takes to access a section of sequential locations until it is accessed again in a cyclic fashion [3,4].

Figure 2 shows the concept of Multi-Level Orbital Data. The whole memory spins in ORBIT0, which has the longest cycle time. Each memory location is accessed and is available to the outside bus ORBIT0 for one bus cycle. Other sections of the memory are rotating their contents in parallel at the same time in a cyclic fashion in each orbit with different cycle time.

Multiple of parallel ORBIT1 each has a portion of the memory spinning at a faster cycle time than ORBIT0 because it contains a smaller number of memory locations. So Multiple of parallel ORBIT1 form ORBIT0 and multiple of parallel ORBIT2 form one ORBIT1.

There is no extra memory storage or duplication of data in any orbit. This is because the whole memory is divided to an integer number of sections for ORBIT1 and ORBIT2. There are multiple ORBIT2 rotating in parallel at the fastest orbit cycle and multiple of ORBIT1, each rotating in parallel at a slower cycle as each ORBIT1 consists of the contents of multiple ORBIT2. ORBIT0 rotates at the slowest cycle as it consists of all data in all lower-level orbits.

For example, if each ORBIT2 has N data elements, its orbital period will be N cycles and if there is I number of parallel ORBIT2, then ORBIT1 will have $N \times I$ data and its orbital period is $N \times I$ cycles. If there is K parallel ORBIT1, the ORBIT0 will have $N \times I \times K$ data elements and its orbital period will be $N \times I \times K$ cycles.

We presented Parallel Vector Processing Using Multi Level Orbital Data [5]. The data move in different orbits to become available to other processors in higher orbits at different time. We use this memory to apply parallel vector operations to data streams at first orbit level. Data processed in the first level move to upper orbit one data element at a time, allowing a processor in that orbit to apply another vector operation to deal with serial code limitations inherited in all parallel applications and interleaved it with lower-level vector operations.

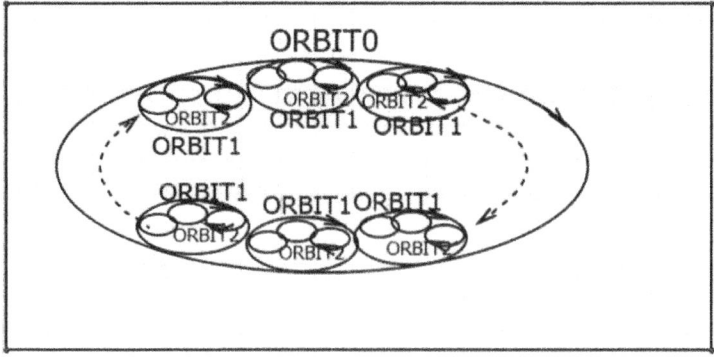

Fig. 2. Concept of Multi-Level Orbital Data

4   MIMD Model for Parallel Computing with Orbital Data

Figure 3 shows a block diagram of a parallel computing using orbital data. Multiple parallel orbits in the same level are used to map different data in parallel orbits at the same speed and have the same orbital period. Parallel orbits orbitk0, orbitk1, ..and orbitkn have different data and all supply data at the same time to be accessed using different instructions.

For example: orbitk0 data D0 is delivered to P0 at time T0 and uses instruction I0 to process it, at the same time data D1 is supplied to P1 in orbitk1 and uses instruction I1 to process it, and Dn is supplied to Pn at T0 and uses instruction In to process it. Instructions are delivered from an instruction pool or a parallel orbit.

All processors P0, P1,..Pn ,at the same time, are executing different instructions while they are processing different data. The time is known to all processors so that in a specific time, each processor executes the specific instruction to process its data in parallel. This MIMD model is useful for parallel computing applications.

Data sharing could be used among different processors in a higher orbit as each data element in these parallel orbits are combined and appear each at specific time in the higher orbit as explained in next section.

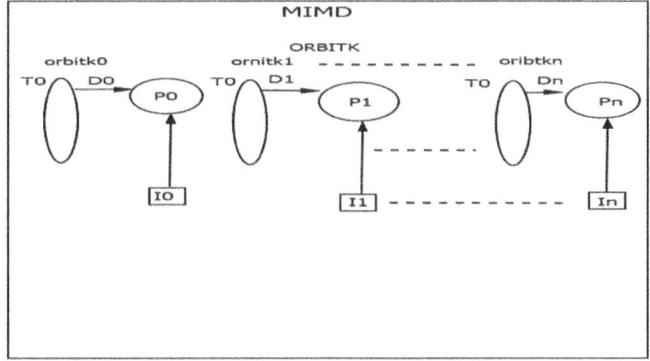

Fig. 3. Parallel Computing with Orbital Data Model

4.1   Explaining the Operations of the Orbital Data Parallel Computing

Figure 4 shows different processors accessing different data sections at the same time in different orbits. P3, P4, P5 and P6 processors connected to multiple parallel orbits, as in ORBIT2 in Fig. 2, and accessing X1..Xn, Y1..Yn, W1..Wn and Z1 .. ZN data elements at same time 1 to N. Each processor is connected to one of ORBIT2 and accesses the same data in a cyclic fashion every N cycles. So for example P3, access X1..Xn from time cycles 1 to N and N to 2N , ..and this repeats every N cycle.

Processors P1 and P2 are connected to multiple parallel higher-level orbits as in ORBIT1 of Fig. 2. P2 is accessing the data elements Y1...Yn, followed by Z1..ZN at time 1 to 2N and at the same time P1 is accessing data X1..Xn, followed by W1..Wn. This repeats every 2N cycle. In the highest level, ORBIT0, P0 access the whole data X1..Xn, followed by Y1..Yn followed by W1..Wn followed by Z1..Zn in time from 1to 4N.

4.2   Parallel Operations

– P0 in ORBIT0 accesses X1..Xn in time 1..N and at the same time P1 in ORBIT1 accesses X1..Xn and P3 in ORBIT2 accesses X1..Xn .
– P0 in ORBIT0 accesses Y1..Yn in time N..2N and at the same time P4 in ORBIT2 access Y1..Yn.
– P1 in ORBIT1 accesses X1..Xn in time 2N..3N and at the same time P3 in ORBIT2 access X1..Xn.
– P0 in ORBIT0 accesses Z1..Zn in time 3N..4N and at the same time P2 in ORBIT1 access Z1..Zn and P6 in ORBIT2 access Z1..Zn.

4.3   Interleaving Operations

While the processors are accessing data in the same time, the same processors can access different data in interleaved time as:

– P1 and P3 share X1..XN and can access them as:
  P1 accesses W1..WN in time N..2N while P3 access X1..XN in time N..2N, then P1 accesses X1..XN later in time 2N..3N to continue the work on the same data used by P3 before without the need to transfer data using messages or synchronization.
– P0 and P5 share W1..WN and both can access them as:
  P0 accesses Y1..YN in time N..2N while P5 accesses W1..WN in time N..2N, then P0 accesses W1..WN later in time 2N..3N in interleaved fashion.
– P2 and P6 share Z1..ZN and both can access them as:
  P2 accesses Y1..YN in time 1..N while P6 access Z1..ZN in time 1..N, then P2 accesses Z1..ZN later in time N..2N to continue processing the same data used by P6 before.

5   The Implementation of a Parallel Computing System Using Orbital Data

Figure 5 shows a multi-level Orbital Data Memory consists of multiple sections, each section rotates its content around a special bus as shown for DRAM banks BNK0, BNKM, BNKN.

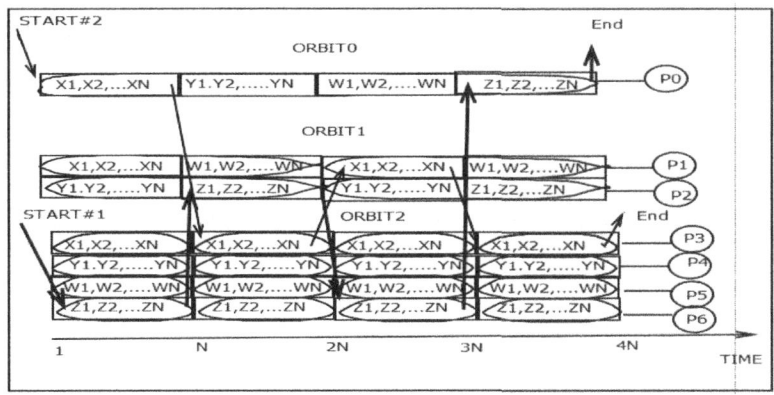

Fig. 4. Operations of Orbital Data Computing

The bus does not need any address lines or arbitration signals, only a clock, start signal and the data [3,4]. The bus for each bank has all locations continuously spinning at cycle time equal the number of locations in the bank multiplied by bus clock time. The system combines several numbers of banks and makes their contents available one after another in sequence using multiplexers and demultiplexers. The bus output of multiplexer MUX1 represents ORBIT1 in Fig. 2. and is connected as input for demultiplexer De-MUX1. If MUX1 is selecting BNK3, then the data out from BNK3 is delivered to MUX1 bus and rerouted through De-MUX1 to be stored back in BNK3 storage, while other banks data outputs are connected to their banks to deliver their data back at the same time in parallel representing ORBIT2 in Fig. 2.

MUX0 is used to access the whole memory at level 0 and represents ORBIT0 in Fig. 2., by selecting, in sequential order, the outputs from all MUX1. The bus of MUX0 is also connected to the input of De-MUX0 and rerouted to be stored in the same accessed bank through DE-MUX1.

Figure 5 also shows a multiprocessor organization using Multi-Level Orbital Data. Each group of processors is connected to one multiplexer/De-multiplexer to access sections of the memory. Other groups of processors are similarly connected to other levels of Orbital data. Each processor group shares one portion of the memory each at a specific time in parallel with the other groups of processors. One group shares one multiplexer/De-multiplexer MUX0, DE-MUX0 that represents the highest level ORBIT0 as P0 in Fig. 4. The second group shares MUX1/DE-MUX1, that represents ORBIT1 as P1, P2 in Fig. 4., accessing a smaller portion of memory that spins at higher speed. The lowest first level of processors are connected directly to the memory banks or sections that each represents ORBIT2 as P3, P4, P5, P6 in Fig. 4.

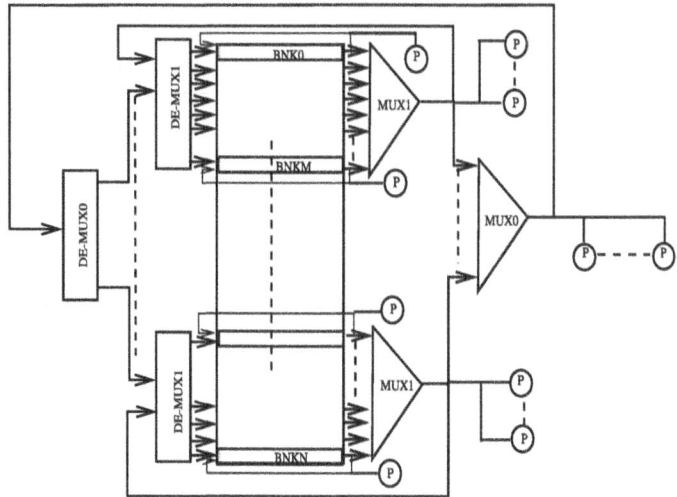

Fig. 5. Implementation of Orbital Data Computing

5.1   Read and Write Operations for the Orbital Data Based Parallel Computing

For the system given above in Fig. 5., each orbital data becomes available to the processor in the bus at a specific time that is mapped to it and can read it and is also available to multiple other processors in the different orbits at the same time.

When a data element is provided to a processor in a higher-level orbit through the multiplexer, the right to write to this element is given to the processor in the higher orbit. The frequency of data in a lower orbit is multiple times that of a higher orbit because the higher orbit combines multiple of the parallel lower orbits. To explain, for example in Fig. 4, the data element X1..XN occurs four times from 1 to 4N in ORBIT2 and two times in ORBIT1 and one time in ORBIT0. So, if P0, P1, and P3 are requesting to write to X1, the priority is given to P0 because P3 will have the opportunity to write soon in the next N cycles and P1 to write after 2N cycles to write to X1.

6   Performance Evaluation of the Orbital Data Parallel Computing

We assume P processors to solve the finite element analysis application using conventional parallel processing and compare it to the orbital data based parallel computing. The cost of computation for a single processor assuming a matrix $N \times N$ to get the average of 4 neighboring elements plus the element itself [2] is $N \times N$. The cost of communication for each process $= 2N =$ size of two boundary rows when it uses strip decomposition to update data in each process [2].

6.1   The Performance of Conventional Parallel Computing

Conventional Parallel Computing using a single shared bus must wait for all processors one at a time to update the neighboring rows, therefore the communication cost is equal to $2 \times N \times P$ [2].

The cost of computation is $\frac{N \times N}{P}$ [2].

Total cost per processor = computation time + communication time= $2NP + \frac{N \times N}{P}$

The Scalability is measured compared to the time it takes for performing the application using single processor to the time of parallel computing as: -

Scalability = $N \times N$ divided by the time using parallel computing as:

$$Scalability = \frac{N \times N}{\frac{N \times N}{P} + 2NP} \tag{1}$$

So as P increases the cost of communication increases limiting the scalability of large number of processors for the conventional parallel computing.

6.2   Performance of Orbital Data Parallel Computing

We assume that in the first level orbits, processors are responsible for computation.

The second level orbits processors are responsible for communication between neighboring processes.

The first level orbits divide the data equally among P orbits; therefore, each first level orbit has NxN divided by P of the data elements. The computation time in each of first level orbit is therefore $\frac{N \times N}{P}$.

Each element is updated by writing the new average to its data by the processor one element at a time. The second level orbits pass the data between neighboring processes. The cost of communication is equal to 3N (last row of neighboring process+ shared boarder row + second row of the current process).

Total cost per processor = computation time + communication time= $3N + \frac{N \times N}{P}$

The scalability of orbital data computing is the time of single processor divided by the time of orbital data-based computing as:

$$Scalability = \frac{N \times N}{\frac{N \times N}{P} + 3N} \tag{2}$$

6.3   The Results

We assume an application matrix that has NXN= 1000×1000 data elements and change the number of processors to calculate the scalability for the conventional parallel computing and orbital data-based computing using the above equations.

Figure 6 shows the scalability of the conventional parallel computing and the orbital data-based computing for the equation solver application of 1000×1000 matrix versus the number of processors.

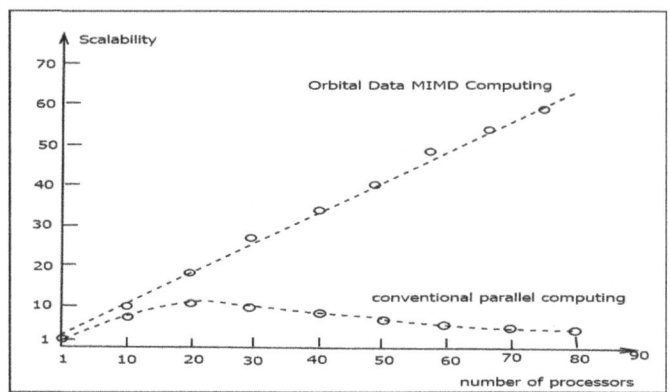

Fig. 6. Scalability of Conventional parallel Computing and Orbital Data Computing

The scalability of the orbital-data based computing is linearly proportional to the number of processors because N is much larger than P and the cost of communication is independent on the number of processors. The interleaving of communication with computation time is accomplished with the use of the higher-level orbit, as explained above, could reduce the overhead of communication that is the main reason for limiting the scalability in parallel computing.

For the conventional parallel computing the cost of communication is linearly proportional to the number of processors causing its scalability to be limited to a small number of processors. The scalability of the novel orbital data computing is linearly proportional to the number of processors which is essential to solve big data size applications and fully utilizes the improvements in technology that allows the implementation of large number of processors in a single chip.

7 Conclusions and Future Work

Orbital data allows the transfer of data to different orbit levels at the same time with different cycle time. This allows processors to overlap their work and hide dependencies, and fully utilizes processor time. It improves the performance of a large number of processors working together in parallel and also deals with the fundamental scalability problem that faces conventional parallel computing when accessing shared data.

Future work will include the development of this new computing and the implementation in FPGA also the development of software and algorithms.

References

1. Hennessy, J., Patterson, D.A.: Computer Architecture: A Quantitative Approach. Morgan Kaufmann Publishers Inc, San Francisco, CA (1996)
2. David E. Culler, Jaswinder Pal Singh, with Anoop Gupta: Parallel Computer Architecture: A Software/Hardware Approach. Morgan Kaufmann Publishers, San Francisco, California ISBN 1-55860-343-3
3. Mekhiel, N.: Data processing with time-based memory access, US 8914612B2 Dec 16 (2014)
4. Mekhiel, N.: Introducing TAM: Time Based Access Memory. IEEE Access journal, SPECIAL SECTION ON SECURITY AND RELIABILITY AWARE SYSTEM DESIGN FOR MOBILE COMPUTING DEVICES, March 30, 2016.P1061-1073 Volume 4
5. Mekhiel, N.: Parallel Vector Processing Using Multi Level Orbital Data. International Journal of Computer, Electrical, Automation, Control and Information Engineering Vol:11, No:3, 2017 paper
6. Burger, D., Goodman, J.R., Kägi, A.: Memory bandwidth limitations of future microprocessors. Proc. 23rd Annu. Int. Symp. Comput. Archit., New York, NY, USA, 78–89 (1996)
7. Wulf, W.A., McKee, S.A.: Hitting the memory wall: Implications of the obvious. ACM SIGARCH Comput. Archit. News **23**(1), 20–24 (1995)

Data Processing and Image Analysis

Image Segmentation Network Based on Convolutional Layers, a Clustering Layer and Cluster Classification

Roberto Rosas-Romero[✉]

Universidad de las Américas Puebla, San Andrés Cholula Puebla, México
roberto.rosas@udlap.mx

Abstract. This work contributes to the detection of regions of interest on images and their corresponding classification in medical imaging applications by introducing an image segmentation network that consists of four stages. In the first stage, multi-resolution processing is applied to outline regions where further segmentation and classification are to be conducted. Subsequently, a quad-tree division stage followed by a clustering stage deliver an image that is divided into unlabeled clusters. The output stage assigns each cluster to one class. This architecture offers flexibility in the input and output stages since this network can be fed with images of any size and the output stage can be implemented with any traditional classification model such as *k-nearest neighbors*, *multi-layer perceptron*, and *support vector machine*. Another contribution of this work is that this network does not rely on a very large number of annotated images for its training.

Keywords: Image segmentation · Convolutional Neural Networks · Multi-Resolution Analysis · Clustering

1 Introduction

Image segmentation has been a useful tool in various computer vision applications such as autonomous vehicles, computer assisted surgery and computer-assisted diagnosis [1]. The emergence and updates of convolutional neural networks (CNN) have contributed to advance image segmentation tasks. However, there are some drawbacks such as the need for very large sets of annotated images and classification models with very high computational complexity, which leads to the research of image segmentation approaches that rely on traditional feature extraction and machine learning techniques.

The architecture of the proposed image segmentation network presents similarities and differences with respect to the basic architecture for image segmentation that consists of a deep convolutional encoder-decoder [2] that broadly consists of two stages, the encoder that extracts features from the image through filters, and the decoder that generates a segmentation mask containing the object outline. Most image segmentation networks have this architecture or a variant of it such as U-Net [3], FastFCN [4], Gated Shape CNN [5], DeepLab [6], Mask RCNN [7]. The training of deep networks

G. Hu et al. (Eds.): CAINE 2024, CCIS 2242, pp. 167–175, 2025.
https://doi.org/10.1007/978-3-031-76273-4_13

for image segmentation requires a very large number of annotated images and data augmentation to increase the number of available annotated images. We present a network originally developed for segmentation of biomedical images that does not rely on the strong use of available annotated images and data augmentation such as the detection and classification of skin burns on color images.

The fundamental operation in deep convolutional networks is the *convolution* [8], also known as *filtering*, under the frequency domain, where the implementation of convolution relies on sliding a *kernel* $k(m, n)$ over the image to be segmented $I(m, n)$ to generate a filtered version of the input image, highlighting certain features. After convolving the input image with the sliding kernel, a *pooling* operation is used to reduce the dimensions of the filtered image and keep significant information. At the convolutional layers, the kernel entries are the parameters to be learned. Since the number of layers in deep CNNs is large, the number of kernel parameters to be learned is huge.

In the proposed approach, the implementation of convolutional encoding-decoding is developed with the *discrete wavelet transform* (DWT) and the *inverse discrete wavelet transform* (IDWT) [9]. Thus, the kernels of convolutional layers are wavelets, which correspond to filters that detect specific features related to texture and intensity variations. Both, the proposed approach and deep CNNs use the convolution; however, kernels in deep CNNs are learned while wavelets are predefined, which is an advantage in terms of computational complexity. In deep CNNs, a *Rectified Linear Unit* (ReLU) is used to learn non-linear features after convolution. In the proposed approach, after encoding-decoding (DWT-IDWT), quad tree-division and a threshold activation function $E > T$ are used to generate a mask that outlines the object of interest where further and detailed segmentation is to be conducted. Another advantage of using DWT-IDWT for image segmentation is its flexibility in terms of the dimensions of the image to be processed. To learn the kernel parameters in deep CNNs, the size of the images under segmentation has to be predefined before training.

Another difference between the proposed approach and deep CNNs takes place in the output layers, where semantic segmentation is conducted. Deep CNNs classify image patches into distinct clusters through encoding-decoding. Encoding consists of convolutional and pooling layers, which gradually reduce image dimensions while increasing the number of channels. Decoding consists of up sampling layers, which gradually increase image size while reducing the number of channels. In the proposed approach, the architecture of the layer for semantic segmentation is not deep, but instead consists of a *self-organizing map* (SOM) that groups similar image patches together based on their features. Each neuron in the SOM corresponds to a cluster or group of image patches, and the features of each cluster are fed to classifier that assigns one class. Classification layers in deep CNNs are typically fully-connected. The architecture of the output layers in the proposed approach is flexible since the classification of clusters could be implemented using a classic learning machine (such as a *multilayer perceptron, support vector machine, k - nearest neighbors*), which does not require very large sets of annotated images.

2 Method

The proposed learning image-segmentation approach, shown in Fig. 1, broadly consists of four stages: (1) an encoding-decoding stage based on the multi-resolution analysis that produces an energy map; (2) an energy-based quad-tree division stage, which generates an image mask that outlines the regions of interest where further segmentation is to be conducted; (3) a clustering stage that separates the region of interest into clusters; and (4) a cluster classification stage. The input image is a gray scale image $I(x, y)$ that is the result of combining the RGB channels of a color image with the *luminosity model*, $I(x, y) = 0.21I_R(x, y) + 0.72I_G(x, y) + 0.07I_B(x, y)$ since this model resembles human vision perception, which is more sensitive to green than to other colors.

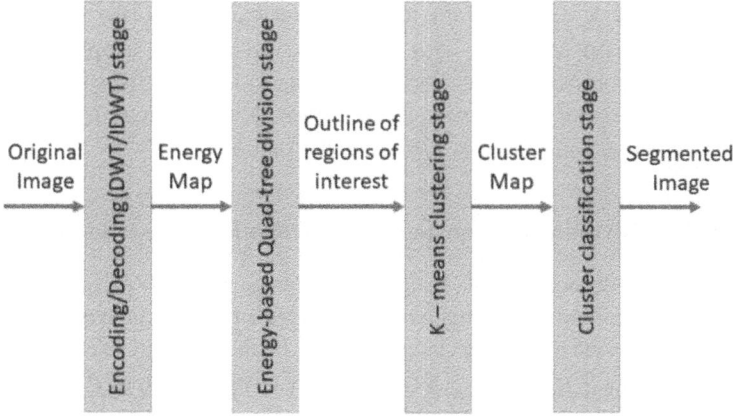

Fig. 1. Proposed approach based on four stages.

2.1 Encoding-Decoding Layers

The encoding-decoding stage consists of convolutional layers that perform a multi-resolution analysis of the image of interest to identify sections where there is significant energy variability [8], discarding homogeneous regions with low energy content. The multi-resolution analysis decomposes an image $I(x, y)$ into four components; a scaling component $\phi(x, y)$ and three wavelet components that measure image intensity variations along columns $\psi^H(x, y)$, rows $\psi^V(x, y)$, and diagonals $\psi^D(x, y)$. The reconstruction of an $M \times N$ image $I(x, y)$, using compressed and spatially shifted versions of these components, is known as the *inverse discrete wavelet transform* (IDWT) [9],

$$I(x, y) = \frac{1}{\sqrt{MN}} \sum_m \sum_n W_\phi(m, n)\phi(2x - m, 2y - n) + \frac{1}{\sqrt{MN}} \sum_m \sum_n \sum_{i \in \{H, V, D\}} W_\psi^i(m.n)\psi^i(2x - m, 2y - n)$$

$$(1)$$

where the coefficients that multiply the components $W_\phi(m, n)$, $W_\psi^H(m, n)$, $W_\psi^V(m, n)$, $W_\psi^D(m, n)$ are the *discrete wavelet transform* (DWT) of image $I(x, y)$,

$$W_\phi(m, n) = \frac{1}{\sqrt{MN}} \sum_m \sum_n I(x, y)\phi(2x - m, 2y - n) \tag{2a}$$

$$W_\psi^i(m, n) = \frac{1}{\sqrt{MN}} \sum_m \sum_n I(x, y)\psi^i(2x - m, 2y - n); i = H, V, D \tag{2b}$$

The DWT of $I(x, y)$ (Eq. 1) can be computed as the one-dimensional DWT of the rows followed by the one-dimensional DWT of the columns of $I(x, w)$ since the scaling and wavelet components can be separated: $\phi(2x - m, 2y - n) = \phi(2x - m)\phi(2y - n)$. Thus, computation of the DWT can be described as filtering rows and columns with the impulse response (a. k. a. kernel) of a low-pass filter $h_\phi(n) = \frac{1}{2}[\delta(n) + \delta(n - 1)]$ and a high-pass filter $h_\psi(n) = \frac{1}{2}[\delta(n) - \delta(n - 1)]$, followed by down-sampling. The output of this filter bank consists of four quarter-size images: the *approximation coefficients* $W_\phi(m, n)$, the *horizontal detail coefficients* $W_\psi^H(m, n)$, the *vertical detail coefficients* $W_\psi^V(m, n)$, and the *diagonal detail coefficients* $W_\psi^D(m, n)$.

2.2 Quad-Tree Division based on an Energy Activation Threshold Function

After the encoding-decoding stage, the generated image $\hat{I}(x, y)$ is partitioned into four rectangles and, depending on rectangle energy, each rectangle is recursively split into four rectangles. The criterion to split a rectangle depends on the rectangle energy being above a threshold according to $E_i = \sum_{x \in R_i} \sum_{x \in R_i} \hat{I}^2(x, y) > T_i$, where $0.05E < T_i < 0.18E$, and E is the energy of the rectangle parent. Thus, the image is partitioned into rectangles based on a quad-tree division during two or three iterations if the rectangle energy is significant. After quad-tree division, the selection of significant-energy rectangles results in an image mask $M(x, y)$, where segmentation is going to be conducted.

2.3 Clustering Stage

At the clustering stage, *k-means* [10] is conducted on the masked image $M(x, y)I(x, y)$ instead of the full image $I(x, y)$. Since k-means is an iterative process, this layer operates as a self-organizing map. Quad-tree division provides k - means with the initial means, which are the centers of the rectangles generated after division. In addition, quad-tree division identifies the number of clusters, which corresponds to the number of rectangles. During k -means, the ith cluster is characterized by a feature vector m_i. Given an initial set of k clusters, with means/seeds $\{m_1, m_2, \ldots, m_k\}$, clustering alternates between two steps: *assignment* and *update*. During assignment, an image patch p is assigned a cluster C_i by (1) extracting a feature vector x from a larger patch, and (2) finding the closest cluster mean according to $C = \underset{C_i}{argmin}\|x - m_i\|_2$. The dimensions of the patch used for feature extraction are larger than the dimensions of the patch, to be assigned to a cluster. During update, the cluster means $\{m_i; i = 1, 2, \ldots, k\}$ are adjusted by considering the feature vectors of the patches assigned to that cluster

$\{x_j \in C_i\}$, according to $m_i = \frac{1}{n_i} \sum_{x_j \in C_i} x_j$. Features for clustering are: (1) the normalized centroid coordinates $\left(\tilde{r} = \frac{r}{rows}, \tilde{c} = \frac{c}{columns}\right)$, (2) the normalized average RGB values $\left(\tilde{\mu}_R = \frac{\mu_R}{255}, \tilde{\mu}_G = \frac{\mu_G}{255}, \tilde{\mu}_B = \frac{\mu_B}{255}\right)$, (3) the normalized variance of the RGB values $\left(\tilde{\sigma}_R^2 = \frac{\sigma_R^2}{255}, \tilde{\sigma}_G^2 = \frac{\sigma_G^2}{255}, \tilde{\sigma}_B^2 = \frac{\sigma_B^2}{255}\right)$, (4) the average value of the normalized luminosity of the patch, (5) the average value of the red to green ratio of the patch, (6) texture features. To extract texture features, the *gray level co-occurrence matrix* (GLCM) of a cluster is generated. The GLCM is a matrix that measures the occurrences of a pair of pixel intensity values (I_m, I_n) at distance d and orientation θ [11], according to

$$P(I_m, I_n, d, \theta) = \frac{Pixel\ pairs\ at\ distance\ d\ and\ angle\ \theta}{Tota,\ number\ of\ pixel\ pairs} \tag{3}$$

where the GLCM was generated for four angles ($\theta = 0°, 45°, 90°, 135°$) and various distance values (d = 3, 5, 7, 9). There are 28 Haralick texture features derived from the GLCM, but the used textural features are the *angular second moment, contrast, inverse different moment, entropy,* and *correlation,*

$$ASM = \sum_{m=1}^{L} \sum_{n=1}^{L} P(I_m, I_n, d, \theta)^2 \tag{4}$$

$$Contrast = \sum_{k=1}^{L} k^2 \left[\sum_{m=1}^{L} \sum_{\substack{n=1 \\ |I_m - I_n| < L}}^{L} P(I_m, I_n, d, \theta)^2 \right] \tag{5}$$

$$IDF = \sum_{m=1}^{L} \sum_{n=1}^{L} \frac{P(I_m, I_n, d, \theta)}{1 + (I_m - I_n)^2} \tag{6}$$

$$H = -\sum_{m=1}^{L} \sum_{n=1}^{L} P(I_m, I_n, d, \theta) log_2 P(I_m, I_n, d, \theta) \tag{7}$$

$$Corr = \frac{\sum_{m=1}^{L} \sum_{n=1}^{L} I_m I_n P(I_m, I_n, d, \theta) - \mu_m \mu_n}{\sigma_m \sigma_n} \tag{8}$$

The total number of extracted textural features from each cluster is *4 angle values × 4 distance values × 5 Haralick features* = 80. The total number of intensity and textural extracted features from each cluster is 10 + 80 = 90.

2.4 Cluster Classification

At the output stage, each cluster, generated with k-means, is classified by extracting features and using a trained supervised learning machine, which could be *k nearest neighbors* (KNN), *multi-layer perceptron* (MLP), or *support vector machine* (SVM). The set of features that are used to classify a cluster includes almost all the features that are used for clustering without the use of the normalized centroid positions.

3 Results and Discussion

To test the image segmentation network, experiments were conducted on 30 color images to outline skin burns and classify the burn degree. The image segmentation network was trained to identify six classes; *healthy skin, first-degree burn, second-degree burn, third-degree burn, shadowed skin, background*. Two implementations of the output layer for cluster classification were tested, *k-nearest neighbors* (k-NN) and *multi-layer perceptron* (MLP). The testing image file formats were JPG and PNG, with resolutions ranging from 12.13 pixels per inch to 89.55 pixels per inch. Cluster classification had to be invariant to cluster size since clusters were characterized by irregular shapes of different area sizes. Thus, 6,000 image patches of various sizes (9×9, 11×11,..., 23×23) were used to train classifiers. The training set was obtained from 114 images publicly available and generated by the Biomedical Image Processing Group at University of Seville and Virgen del Rocío Hospital (Seville, SPAIN) [12]. This public data set contains samples from three classes, corresponding to the three skin burn degrees, and does not include samples for healthy skin, shadowed skin, and background. Thus, this data set was complemented with samples from 22 images to introduce patches for the other three classes.

Figure 2 shows the image processing conducted on a color image (Panel A) over the four stages of the proposed network architecture. Panel B shows the result of the DWT-IDWT (encoding-decoding) stage, which is an energy map. Panel C shows the result of the second stage, which is quad-tree division based on a threshold activation function $E_i > T_i$, where the threshold for region T_i is defined as $T_i = \alpha E$, being E the energy of the parent rectangle, and α a parameter manually adjusted within a range of values, $\alpha \in [0.05, 0.18]$. The mask is a union of non-overlapping rectangles characterized by high energy variations. The result of energy-based quad-tree division is a mask that outlines skin regions, including burn injuries. Quad-tree division can be repeated up to three times so that the maximum possible image division can generate up to 4^4 rectangles. Quad-three division also provides the next stage (k-means clustering) with the rectangle centers as the initial seeds for clustering. Panel D shows the result of the third stage, where k-means clustering is conducted. The result of the last stage for cluster classification is shown in Panel E, where clusters were classified as first-, second-, and third-degree burns and colored in red, green, and blue, respectively. Panels E shows the result of cluster classification using MLP while the result of cluster classification based on KNN is shown in Panel F.

The burn detection and classification performance is shown in Table 1. The *sensitivity* and *precision* metrics were measured for each burn degree (first, second and third degrees) and for each classifier (k-NN and MLP). *Sensitivity* is the ratio of detected burns over total number of burns, $Sensitivity = \frac{TP}{TP+FN}$, where TP (True Positives) is the number of correctly classified burns and FN (False Negatives) are non-identified burns. *Precision* is the ratio of correctly identified burns over the total number of identified burns, $Precision = \frac{100 \times TP}{TP+FP}$, where FP (False Positives) are regions incorrectly identified as burns. The last four rows of the Table show the overall burn detection performance that accounts for all burn degrees. In addition, the best classification performance is obtained for third-degree burns since it is easy to discriminate third-degree skin burns from healthy, second- and third-degree burnt skin in terms of intensity and contrast.

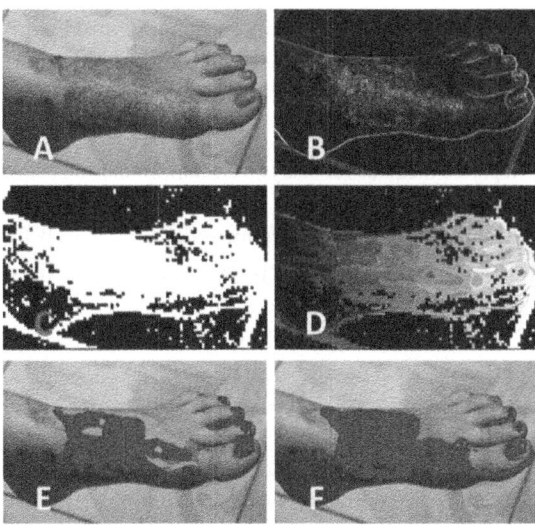

Fig. 2. Detection and classification of skin burns with the image segmentation network at each stage/layer: A) Original image, B) encoding-decoding, C) quad-tree division based on an energy activation function, D) clustering, E) cluster classification using a MLP, F) cluster classification using a KNN.

Table 1. Performance of the proposed method for the detection and classification of skin.

Degree	Classifier	TP	FN	FP	Sensitivity	Precision
First	KNN	30	3	4	0.9091	0.8824
	MLP	31	2	3	0.9394	0.9118
Second	KNN	23	3	3	0.8846	0.8846
	MLP	24	2	3	0.9231	0.8889
Third	KNN	52	4	3	0.9286	0.9455
	MLP	53	3	3	0.9464	0.9464

We have previously worked on detection and classification of skin burns, where each image patch is assigned to a class representing a burn degree by using the sparse representation of feature vectors extracted from image patches [12]. Each feature vector was represented as a linear combination of atoms in over-redundant dictionaries. Image segmentation based on redundant dictionaries was more computationally demanding than the current segmentation approach. The faster classification model was the SVM followed by the MLP.

Table 2 shows different methods for segmentation of skin wounds on color images with the corresponding frameworks and results. This comparison considers the use of different data sets, implementations and experimental setups. Despite data set differences, skin wound recognition required similar image processing and feature extraction techniques. Most of the skin wound recognition methods used similar classifiers and their performance was evaluated using similar metrics, where accuracy and precision were the most common metrics.

Table 2. Reported methods for analysis on skin wounds on color images.

Author	Objective	Framework	Performance
Yadav et al. [12]	Human burn diagnosis	CIELAB color space, support vector machine	Accuracy of 82.43%
Sevik et al. [14]	Classification of skin burns	Texture, fuzzy c-means, multi-layer perceptron	F-score of 80%
Rehman et al. [15]	Segmentation and classification of skin burns	Texture, Otsu's method, multiple classifiers	Accuracy of 74.86%
Khan et al. [16]	Computer aided diagnosis of skin burns	Otsu's method, convolutional neural network	Accuracy of 79.4%
Rangel-Olvera et al. [13]	Segmentation and classification of skin burns	Textural and intensity feature extraction and sparse representation	Sensitivity of 95.65% and precision of 94.02%
Proposed approach	Segmentation and classification of skin burns	Multi-resolution analysis, clustering, texture, intensity, classifiers	Sensitivity of 94.64 % and precision of 94.64%

4   Conclusion

This work proposes an architecture for image segmentation that consists of four stages: (1) encoding-decoding based on multi-resolution analysis, (2) quad-tree division based on regional energy, (3) clustering, and (4) cluster classification. Deep learning models typically require very large amounts of annotated data for training, especially in medical image segmentation, where acquiring and annotating a large data set might not be feasible. Image processing performed through deep network layers often lacks interpretability, which is crucial in medical imaging. On the other hand, the proposed approach is based on less annotated images and relies on traditional feature extraction and machine learning techniques, which make it easier to understand and explain the ongoing image processing. In addition, the proposed approach is more flexible since there are not constrained on the dimensions of the images to be segmented. The proposed image segmentation approach was applied to detect skin burns on color images and to identify their intensity degree for automatic assistance.

References

1. Nida, M.Z., Musbah, J.A.: Survey on image segmentation techniques. Procedia Comput. Sci. **65**, 797–806 (2015)
2. Badrinarayanan, V., Kendall, A., Cipolla, R.: SegNet: a deep convolutional encoder-decoder architecture for image segmentation. arXiv preprint arXiv:1511.00561 (2015)
3. Ronneberger, O., Fischer, P., Brox, T.: U-Net: convolutional networks for biomedical image segmentation. arXiv preprint arXiv:1505.04597 (2018)
4. Wu, H., Zhang, J., Huang, K., Liang, K., Yu, Y.: FastFCN: rethinking dilated convolution in the backbone for semantic segmentation. arXiv preprint arXiv:1903.11816 (2019)
5. Takikawa, T., Acuna, D., Jampani, V., Fidler, S.: Gated-SCNN: gated shape CNNs for semantic segmentation. arXiv preprint arXiv:1907.05740 (2019)
6. Chen, L.C., Papandreou, G., Kokkinos, I., Murphy, K., Yuille, A.L.: DeepLab: semantic image segmentation with deep convolutional nets, atrous convolution, and fully connected CRFs. arXiv preprint arXiv:1606.00915 (2016)
7. He, K., Gkioxari, G., Dollár, P., Girshick, R.B.: Mask R-CNN. arXiv preprint arXiv:1703.06870 (2017)
8. LeCun, Y., Bengio, Y., Hinton, G.: Deep learning. Nature **521**, 436–444 (2018)
9. Mallat, S.G.: A theory for multiresolution signal decomposition: the wavelet decomposition. IEEE Trans. Pattern Anal. Mach. Intell. **11**(7), 674–693 (1989)
10. Dhanachandra, N., Manglem, K., Chanu, Y.J.: Image segmentation using K-means clustering algorithm and subtractive clustering algorithm. Procedia Comput. Sci. **54**, 764–771 (2015)
11. Haralick, R.M., Shanmugam, K., Distein, I.: Textural features for image classification. IEEE Trans. Syst. Man Cybernet. **6**, 610–621 (1973)
12. Yadav, D.P., Sharma, A., Singh, M., Goyal, A.: Feature extraction based machine learning for human burn diagnosis from burn images. IEEE J. Transl. Eng. Health Med. **54**, 764–771 (2015)
13. Rangel-Olvera, B., Rosas-Romero, R.: Detection and classification of burnt skin via sparse representation of signal by over-redundant dictionaries. Comput. Biol. Med. **132**, 1–9 (2021)
14. Sevik, U., Karakullukçu, E., Berber, T., Akbas, Y., Türkyılmaz, S.: Automatic classification of skin burn colour images using texture-based feature extraction. IET Image Proc **13**, 2018–2028 (2019)
15. Rehman-Butt, A.U., Ahmad, W., Ashraf, R., Asif, M., Ashraf-Cheema, S.: Computer aided diagnosis (CAD) for segmentation and classification of burnt human skin. In: Proceedings of the 2019 International Conference on Electrical, Communication, and Computer Engineering (ICECCE), pp. 1–5 (2019)
16. Khan, F.A., et al.: Computer-aided diagnosis for burnt skin images using deep convolutional neural network. Multimed. Tools Appl. **79**, 4545–34568 (2020)

Measuring API Usability and the Environment: A YouTube API Case Study

Sultan Alanazy[1,2] and Jeff Tian[1(✉)]

[1] Dept of Computer Science, Southern Methodist University, Dallas, Texas, USA
`{salanazy,tian}@smu.edu`
[2] College of Computer Science and Info Tech, Imam Abdulrahman Bin Faisal University, Dammam, Saudi Arabia

Abstract. Application Programming Interface (API) plays a vital role in Cloud computing. Improving API usability will help application programmers select and adapt the most suitable APIs for their specific needs. By adopting a general framework for usability measurement and analyzing questions and answers on Stack Overflow for YouTube APIs, we define three direct usability metrics: "team response time" to quantify learnability, "net upvote" to measure user satisfaction, and "downvote" to reflect usability problems. In addition, we identify six environmental factors that influence API usability by analyzing the question tags: language, library, browser, operating system, integrated development environment, and device. Our measurement and analysis provide an objective assessment of API usability and its environmental influencing factors.

Keywords: API (Application programming interface) · usability metrics · Stack Overflow · YouTube API · empirical study

1 Introduction

In today's interconnected world, Application Programming Interface (API) has emerged as a critical element for enabling smooth communication between various Cloud applications and services [9]. The primary users of APIs are application programmers who search, read, understand, and integrate appropriate APIs into their systems [7]. The ease of use and integration of APIs significantly influence their adoption by developers. Moreover, user-friendly APIs enhance developer productivity by promoting intuitive interactions, reducing the need for extensive documentation, and encouraging code reuse.

To comprehensively evaluate API usability, we have previously identified eight direct usability measurements: recognizability, learnability, operability, error protection, UI aesthetics, accessibility, retention, and satisfaction [10]. In addition, we have systematically identified influencing factors of API usability, including the design and implementation, the clarity of documentation, the complexity of the API, the diversity of the user population, and the environment in

G. Hu et al. (Eds.): CAINE 2024, CCIS 2242, pp. 176–186, 2025.
https://doi.org/10.1007/978-3-031-76273-4_14

which it is used [9]. Understanding these influencing factors is crucial for design-ing and maintaining APIs that are user-friendly for the target user population and user tasks across various environments.

In this paper, we define specific metrics to measure usability and influence factors for YouTube APIs, one of the most popular and widely used APIs, based on related Q&A data from Stack Overflow. The measurement and analysis results provide an objective usability assessment and link it to various environmental influence factors.

2   Related Work on API Usability and YouTube APIs

The growing importance of API usability in software development, particularly in cloud computing, has been highlighted by numerous studies [4,6–8,10]. The integration of human-computer interaction principles into API design has been advocated to enhance usability [7]. Inadequate documentation has been identified as a significant obstacle for developers [8]. To mitigate this issue, improvements in documentation quality and inclusion of illustrative code examples have been recommended. In addition, a framework has been developed that consists of 12 cognitive dimensions relevant to API usability, offering a comprehensive evalua-tion of API usability at the cognitive level [4]. Despite the availability of models and guidelines to measure API usability, more comprehensive empirical studies are necessary to validate their effectiveness and address the nuanced aspects of API usability in practice [6].

Recently, we developed a comprehensive framework to measure and enhance API usability, which includes a measurement framework and an analysis frame-work [10]. The measurement framework encompasses both direct measurements of API usability and a comprehensive set of indirect measurements of factors that may influence API usability. The analysis framework empirically validates the measurement framework and establishes predictive relationships between its direct and indirect metrics. In a subsequent study, we systematically identified influencing factors by examining the entities and artifacts involved in API devel-opment and usage [9].

Improving API usability involves collecting and analyzing data from various sources. However, measuring API usability through regular usage or testing can be challenging because it is difficult to directly engage with API users, particu-larly in public APIs, and track their interactions in detail. Alternatively, strate-gies for collecting API usage information through crowd-sourced Q&A forums, like Stack Overflow, were used effectively in several empirical studies [1,2]. In our previous studies, we used this approach to examine the usability of APIs [3,9]. We chose YouTube APIs for our case study, one of the most popular and widely used APIs.

YouTube APIs offer many features, with 47 different methods available in 18 libraries. Moreover, YouTube APIs use JSON or XML for communication, making them compatible with any programming language that can handle these data formats. Stack Overflow collaborates with Google to serve as an online support forum for YouTube APIs. For additional support with YouTube APIs beyond the official documentation, Google encourages users to visit Stack Overflow and utilize the tags "YouTube-API" and "YouTube-data-API" to ask questions or search for relevant answers. This practice reflects the widespread reliance of many API users on alternative sources other than official documentation to obtain information about APIs [8].

Users can post their questions or search for answers on Stack Overflow, thereby enriching the APIs online support resources. Stack Overflow offers a voting system for Q&A, which helps users quickly find solutions to their problems. Each question has one or more answers; however, only one answer can be marked as accepted by the user who asked the question. If the question does not have an accepted answer, the most upvoted answer can be treated as a substitute for the accepted answer.

In this study, we continue our examination of API usability and its influencing factors, focusing on concrete usability metrics and environmental influences for YouTube APIs based on related Q&A data from Stack Overflow. We use Stack Exchange Data Explorer (SEDE) to extract all questions and answers related to YouTube APIs from Stack Overflow between 2008 and 2023. The collected data contain 8743 questions and their corresponding answers.

3   Measuring API Usability

Building upon our framework for API usability measurement and analysis [9,10], we analyzed the Q&A data for YouTube APIs on Stack Overflow to identify direct usability metrics and environmental influencing factors. We have identified three direct metrics of API usability:

- **Team response time** (T_r) measures the duration taken by the support team to respond to API-related inquiries, reflecting API learnability.
- **Net upvote for a specific answer** (Λ_a) gauges user satisfaction by considering the net upvotes for that answer (upvotes minus downvotes).
- **Downvotes for a specific answer** (V_a) measure usability problems based on the number of downvotes the answer receives from users.

APIs learnability is one of the major subcharacteristics of usability as it determines how easily users can understand and learn a product or system [5]. The learnability of APIs depends on several factors, such as the quality of documentation, online support, use background, task complexity, and others. When users begin interacting with APIs, they typically start by consulting the official documentation supplied by the API provider. If they encounter challenges or require additional assistance, they often turn to online support forums like Stack Overflow. Here, they can search for similar questions and seek solutions to their queries. If they cannot find the information they need, they can post detailed questions accompanied by specific tags and wait for responses from the support team. The increased waiting time for an answer will directly contribute to an increase in overall learning time. As a variant of learning time, we define

Table 1. Summary Statistics of Direct Usability Metrics

Metric	Min.	Qt.1	Median	Mean	Qt.3	Max.
T_r	1	73	652	135327	7924	6006929
Λ_a	-10	0	1	3.4	2	5391
V_a	0	0	0	0.06	0	15

a metric "Team Response Time" (T_r), which measures the time it takes for the support team to respond to a specific question. T_r is calculated from the available data as:

$$T_r = T_a - T_q$$

where T_a is the timestamp when the support team replies to the question and T_q is the timestamp when the user first asks the question. A higher T_r indicates lower API learnability, as it represents more time required for learning.

User satisfaction with APIs refers to the degree to which user needs are met when a product or system is used in a specific context [5]. This satisfaction encompasses various aspects related to how users feel after interacting with a product, including its documentation and user support. Our second metric, "Net upvote for a specific answer" (Λ_a), directly measures user satisfaction with the answer provided by the API support team as:

$$\Lambda_a = \text{Upvote}_a - \text{Downvote}_a$$

where $Upvote_a$ represents the number of upvotes for a specific answer, and $Downvote_a$ represents the number of downvotes for the same answer, thus reflecting API user's satisfaction based on their votes.

Based on our examination of Stack Overflow, downvotes are more significant than upvotes in measuring satisfaction levels. This difference arises from Stack Overflow's voting protocols, where downvote privileges are reserved for users with a higher reputation threshold of 125 points, compared to the more accessible 15 points required for upvoting. Additionally, the downvote triggered a reputation penalty, underscoring the weight it carries. Further investigation into Stack Overflow posts related to YouTube APIs revealed a noteworthy fluctuation in the ratio of upvotes to downvotes over the years, between 2009 and 2023. This ratio ranged from 13 to 26 upvotes for every downvote, reflecting the varying degrees of user sentiment. Therefore, downvotes can be considered a direct indicator of usability problems, similar to the role of software defects as a direct indicator of general software quality problems. Our third metric, "downvote for a specific answer" (V_a), reflects usability problems based on users downvoting a particular answer as:

$$V_a = \text{Downvote}_a$$

where $Downvote_a$ is the number of downvotes on an answer related to the question asked by API users.

Table 2. Distribution of Downvotes

V_a	0	1	2	3	4	5	6	8	9	11	15
Frequency	8343	324	46	11	7	4	3	1	1	1	1

The measurement results using these three direct usability metrics are summarized in Table 1. The summary statistics, including minimum, maximum, mean, median, first quartile (Qt.1), and third quartile (Qt.3) for each metric, provide insights into their characteristics and variability, and enhance our understanding of their behavior.

The median response time (T_r) for all questions is 652 min, indicating typical promptness to address various API-related queries. The majority of questions were answered within days, suggesting a prompt response time. However, there were instances where answers were considerably delayed, with 25% of them taking 7924 min or more, and a maximum of 6006929 min. In addition, the difference of 7851 min between the 1st quartile (73 min) and the 3rd quartile (7924 min) highlights the variability in response times.

For Λ_a, the mean user satisfaction score is 3.419, indicating moderate satisfaction levels. Analysis of net upvotes reveals that only 114 responses out of 8743 received negative net upvotes, which is a mere 1.3% of the total. In contrast, 3335 responses (38.14%) received 0 net upvotes, while the majority, 5294 responses (60.55%), received positive net upvotes. This distribution shows mostly positive, or at least neutral, sentiment towards the accepted answers to the user questions, although it also indicates some areas that need attention for improvement. Notably, the difference between the 1st quartile (0) and the 3rd quartile (2) for Λ_a indicates relatively low variability in user satisfaction levels.

Regarding V_a, the vast majority, comprising 8185 responses, received no downvotes, accounting for 95%. However, it is noteworthy that despite the minimum, Qt.1, and Qt.3 all being at 0 downvotes, 399 answers received 1 or more downvotes, with the maximum observed being 15. These 399 answers represent a significant number of potential problems in satisfactorily answering user questions or in providing adequate user support. Table 2 provides detailed information on the distribution of V_a, indicating the frequency of downvote values for each individual answer.

4 Environmental Metrics

Upon analyzing the collected data, we employed matching techniques to link each question and its answer to specific environmental factors based on user-provided tags. These tags enable us to identify six environmental factors for characterizing the usage environment of the APIs and the context of each specific question/answer:

1. Programming Language: Refers to the specific language used to interact with APIs.

2. Operating System: The foundational software managing computer hardware resources, facilitating communication between hardware and software components.
3. Library: A collection of pre-written code modules or functions that extend the capabilities of a programming language.
4. Browser: A software application enabling users to access and display internet content, navigate websites, and interact with online resources.
5. Integrated Development Environment (IDE): A software suite that provides comprehensive tools, frameworks, and features for software development, including code editing and debugging.
6. Device: The physical hardware used to interact with and access digital information, such as computers, phones, and other electronic devices.

Each of these environmental factors is represented as a categorical variable, with its values indicating a specific language, operating system, library, browser, IDE, or device used in the context of API interaction, adoption, and integration. To associate each question and its corresponding answer with a specific environmental factor, we developed a matching function that utilizes the question tags provided by API users.

Starting with the language factor, the YouTube API documentation underscores its language-agnostic nature, prioritizing the use of JSON or XML for communication to ensure compatibility with any programming language proficient in handling these data formats. We examine all tags written by YouTube APIs users or application developers to identify the programming language related to each question. We categorize these tags into three types:

– **Uniquely identified language:** This category pertains to tags that distinctly represent a single language. Within this category, there are a total of 3530 questions, accounting for 40% of the dataset.
– **Multiple languages:** This category relates to tags containing two or more programming languages. We found 103 questions with multiple languages in their tags, representing less than 2% of the dataset.
– **Unspecified language:** This category refers to tags labeled as "UL" when no language tags are identified. In total, there are 5104 questions classified under the unspecified language category, which represents 58% of the dataset.

The most prevalent languages in the dataset are JavaScript, PHP, Python, Java, C#, and Swift.

Similarly, we can extract the other five environmental factors from the tags. However, these factors are much less frequently identified by users, as will be quantified in the measurement results later in this section. The values for the corresponding categorical variable representing each environmental factor are given below:

– Three prominent operating systems emerged: iOS, Windows, and Android, signifying their substantial presence within the dataset.
– Noteworthy libraries with significant representation include Auth Library, Client, Axios, cURL, gdata, jQuery, JSON, Lodash, and Reactjs.

- The typical IDEs encountered in our dataset include .NET, Android Studio, Angular, Django, Eclipse, Flask, Jupyter, Kotlin, Laravel, Node.js, Ruby on Rails, Spring, Visual Studio, WordPress, and Xamarin.Forms.
- The primary browsers identified in our dataset are Chrome, Safari, and Firefox.
- The devices identified within our dataset span a wide range, including desktops, laptops, mobile phones (e.g., iPhone, Galaxy, Redmi), tablets (e.g., iPad), gaming consoles (e.g., Xbox, PlayStation), TV streaming devices (e.g., Roku), and various other devices.

Table 3. Distribution of Programming Languages

Language	Frequency
JavaScript	1625
PHP	831
Python	508
Java	386
C#	289
Swift	238
Ruby	102
ActionScript	52
Dart	19
TypeScript	18
Go	10
Total Identified	**4078**
Total Unidentified	**4665**

Next, we examine the measurement results of the six environmental factors, starting with the language factor in Table 3. JavaScript accounts for 40% of all identified languages in the dataset. The predominance of JavaScript usage suggests a heavy reliance on client-side scripting for YouTube APIs integration. Following closely, the presence of Python and Java indicates their popularity in back-end and server-side development, respectively, reflecting a diverse range of applications utilizing YouTube APIs across different programming paradigms. Moreover, the appearance of C# and Swift in the list, both more than 200, suggests a significant usage of mobile app development in projects related to YouTube APIs. C# is commonly associated with the development of Windows-based desktop and mobile applications, while Swift is the primary language for the development of iOS apps.

In the analysis of the IDE factor, iframe emerged as the most frequent, with 384 occurrences, likely used to embed YouTube video players into web layouts. Following closely, OAuth appeared 325 times, likely linked to its integration for security purposes. Additionally, Ajax and XML each showed notable occurrences, with 57 instances each, indicating their roles in dynamic web development and XML data handling, respectively.

In exploring the browser factor, we found that Chrome, Firefox, and Safari are the primary browsers used by developers interacting with YouTube APIs. Chrome emerged as the most popular choice, with 62 occurrences, followed by Firefox with 42 occurrences, and Safari with 38 occurrences.

In examining the operating system factor, Android, iOS, and Windows emerge as the primary operating systems in the data set. Android appears as the most prevalent, appearing 1228 times, indicating its widespread adoption among mobile developers. Following Android, iOS appears with 370 occurrences, while Windows exhibits a lesser but still notable presence, with 45 occurrences among developers utilizing YouTube APIs.

Upon analysis of the device factor, we found that more developers utilize iPhones, with 114 occurrences, followed by the mobile tag and iPad with 34 and 18 occurrences respectively. However, the TV, the phone, and desktop exhibit lower frequencies, with occurrences of 12, 10, and 5, respectively.

In the analysis of the library factor, notable usage of jQuery, JSON, and client libraries among developers is evident, with 447, 228, and 282 occurrences, respectively. Additionally, the GData and React libraries are utilized, with 135 and 33 occurrences, respectively. However, Curl exhibits a lower frequency compared to other libraries, appearing only 41 times.

Upon examination of the total identified and unidentified data points for each factor, the distribution is as follows. For the programming language factor, there are 4078 identified instances (46.64%) and 4665 unidentified instances. For the IDE factor, there are 1137 identified instances (13%) and 7606 unidentified instances. The browser factor includes 142 identified instances (1.62%) and 8601 unidentified instances. The operating system factor contains 1627 identified instances (18.60%) and 7116 unidentified instances. The device factor has 193 identified instances (2.20%) and 8550 unidentified instances. Lastly, the library factor consists of 1,066 identified instances (13.33%) and 7577 unidentified instances.

5   Environment and Usability

Utilizing our established modeling framework [10], we next examine the potential influence of various environmental factors, defined in Sect. 4, on the direct API usability metrics, defined in Sect. 3. The findings linking direct API usability metrics and environmental factors can provide recommendations for optimizing environmental setups for API users, including specific hardware, network settings, and other customizations to enhance the user experience.

Among the six environmental metrics identified in Sect. 4, only the programming language factor can be identified for a significant number (4078) and proportion (46.65%) of the data points. Therefore, we only analyzed the relationship between the language factor and the direct API usability metrics defined in Sect. 3.

Figure 1 gives the T_r distribution summary in boxplots for different sets of questions tagged with the corresponding languages. The first and third quartiles are represented as the lower and upper boundaries of each box, respectively; and the median is represented by the horizontal line within the box. Significant disparities were observed in the median response times (T_r) between Java, JavaScript, and Swift. Specifically, questions tagged with Java are associated with a median response time of 110 min, whereas questions tagged with JavaScript and Swift have much shorter median response times, less than 25 min. This indicates that the learnability of YouTube APIs is higher when application programmers use Swift or JavaScript compared to Java.

Next, we explored the relationship between the language factor and user satisfaction metric (Λ_a). However, our analysis did not reveal any observable pattern between language factor and Λ_a. This implies that factors beyond the scope of our study may exert a more significant influence on Λ_a.

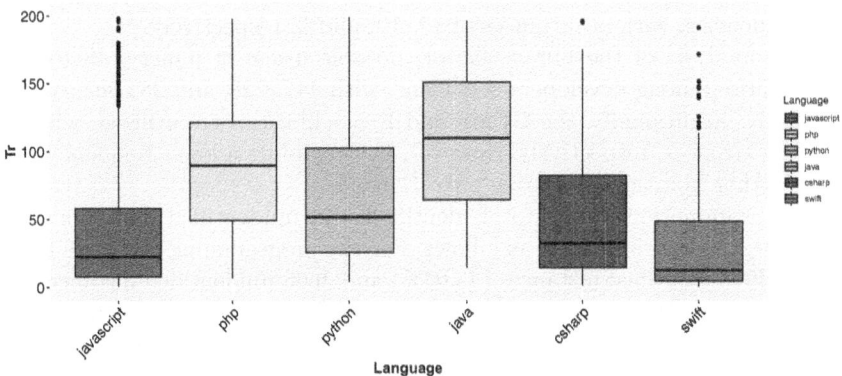

Fig. 1. Comparison of team response time (T_r) across various programming languages

Finally, we investigated the potential impact of language factor on downvotes for a specific answer (V_a). Due to the narrow range of the distribution V_a and the prevalence of data points with $V_a = 0$, we did not find any overall pattern between the type of language used and V_a. We plan to examine this highly skewed distribution for V_a and its relationship to environmental metrics using other analysis techniques more appropriate for this type of data in the future.

6   Conclusions and Perspectives

In this study, we defined a set of concrete usability metrics to provide an objective assessment of API usability from a user's perspective while focusing on the YouTube API as a case study. By adopting a general framework for usability measurement and examining questions and answers related to YouTube APIs on

Stack Overflow, we defined three direct usability metrics: "team response time" to measure API learnability, "net upvote" to measure API user satisfaction, and "downvote" to measure usability problems.

In addition, we identified six environmental factors that influence API usability: programming language, library, browser, operating system, integrated development environment (IDE), and device. Our analysis revealed that the choice of programming language affects API learnability, with Swift and JavaScript users experiencing shorter response times compared to Java users. However, we did not find an observable pattern between programming language and user satisfaction (net upvotes) or usability problems (downvotes), suggesting that other factors may play a more substantial role in these aspects or different analysis techniques are needed to characterize the relationships. These findings highlight the importance of considering environmental factors when selecting and using APIs. Application programmers can make more informed decisions by understanding how different environments impact API usability. API providers can also leverage this information to improve API documentation and support resources, particularly for languages with steeper learning curves.

In future work, we plan to expand our research to include more public APIs and investigate the impact of other influencing factors, beyond the environmental factors examined in this study, on API usability. We will also employ advanced analysis techniques, such as tree-based models, to uncover complex relationships between direct usability metrics and their influencing factors. This comprehensive approach will help us offer practical suggestions for API providers to improve overall API usability and application programmers' experience.

References

1. Ahasanuzzaman, M., Asaduzzaman, M., Roy, C.K., Schneider, K.A.: Classifying stack overflow posts on api issues. In: 2018 IEEE 25th International Conference on Software Analysis, Evolution and Reengineering (SANER), pp. 244–254 (2018)
2. Barua, A., Thomas, S.W., Hassan, A.E.: What are developers talking about? an analysis of topics and trends in stack overflow. Empir. Softw. Eng. **19**(3), 619–654 (2014)
3. Bokhary, A., Tian, J.: Measuring cloud service apis quality and usability. In: Proceedings of the International Conference on Software Engineering Research and Practice (SERP), pp. 208–214 (2018)
4. Clarke, S.J., Becker, C.: Using the cognitive dimensions framework to measure the usability of a class library. In: Annual Workshop of the Psychology of Programming Interest Group (2003)
5. ISO/IEC: Systems and software engineering – Systems and software quality requirements and evaluation (SQuaRE) – System and software quality models. ISO/IEC 25010, International Organization for Standardization, Geneva, Switzerland (2011)
6. Mosqueira-Rey, E., Alonso-Ríos, D., Moret-Bonillo, V., Fernández-Varela, I., Álvarez Estévez, D.: A systematic approach to api usability: Taxonomy-derived criteria and a case study. Inform. Softw. Technol. **97**, 46–63 (2018)

7. Myers, B.A., Stylos, J.: Improving API usability. Commun. ACM **59**(6), 62–69 (2016)
8. Robillard, M.P.: What makes apis hard to learn? answers from developers. IEEE Softw. **26**(6), 27–34 (2009)
9. Tian, J., Alanazy, S., Bokhary, A., Alharthi, S., Ghanem, S.: Measuring influencing factors of api usability. In: 2022 International Conference on Computational Science and Computational Intelligence (CSCI), pp. 1889–1894 (2022)
10. Tian, J., Bokhary, A., Alanazy, S.: A comprehensive framework for measuring and improving api usability. In: 2021 International Conference on Computational Science and Computational Intelligence (CSCI), pp. 1958–1963 (2021)

Networking and Edge Computing

Enabling Seamless Healthcare Services with RC-Based Geographical Peer-to-Peer Fog Architecture in the Healthcare 4.0

Hanumanthu Vinaganti[1], Nikhitha Chowdary Tannedi[1], Pooja Nannuri[1], Indranil Roy[1(✉)], Reshmi Mitra[1], Nick Rahimi[2], and Bidyut Gupta[3]

[1] Southeast Missouri State University, Cape Girardeau, MO 63701, USA
{hvinaganti1s,ntannedi1s,pnannuri1s,iroy,rmitra}@semo.edu
[2] University of Southern Mississippi, Hattiesburg, MS 39406, USA
nick.rahimi@usm.edu
[3] Southern Illinois University Carbondale, Carbondale, IL 62901, USA
bidyut@cs.siu.edu

Abstract. This paper introduces a novel peer-to-peer (P2P) architecture tailored for resource discovery in sensor-based health-monitoring devices. Departing from traditional publish/subscribe methods, this architecture employs location-, interest-, and resource-based mechanisms to enhance efficiency and security without relying on a centralized communication model. The proposed architecture comprises a two-level overlay network: a transit ring housing group-heads representing specific resource types, and a fully connected group of peers. Theoretical analysis demonstrates that search latency remains independent of the number of peers, with constant complexity for intra-group data lookup and $O(n)$ complexity for inter-group data lookup, where n signifies the total number of resource types. This architecture enables efficient, cost-effective, and secure management of large data throughput for medical IoT systems, offering an alternative to the centralized communication model.

1 Introduction

Digital healthcare products, technologies, services, and enterprises have emerged from the adoption of Healthcare 4.0 standards, integrating robotics, artificial intelligence (AI), big data analysis, cloud computing, and real-time actuators. These innovations transform healthcare systems by enabling real-time patient assistance and focusing on early disease detection and prevention.

Despite these benefits, challenges such as communication delays between sensors and clouds, high costs of cloud-based data storage and transfer, and time-sensitive data privacy issues limit the architecture of sensor-reliant healthcare

G. Hu et al. (Eds.): CAINE 2024, CCIS 2242, pp. 189–201, 2025.
https://doi.org/10.1007/978-3-031-76273-4_15

applications. Network outages or congestion can endanger patients' lives by disrupting communication between sensors and cloud services. Numerous studies [6,14] have addressed the security of personal health information.

Fog computing, combined with cloud-based analytics, addresses latency, privacy, and scalability challenges in healthcare data processing. By utilizing edge and networking devices, fog computing reduces latency by processing data closer to sensor nodes. This distributed architecture splits complex tasks into smaller ones, enhancing processing efficiency. Although fog devices have limited resources, they host cloud services or APIs for prompt data processing, improving security, reducing costs, and optimizing data analysis. Challenges such as service availability and infrastructure reorganization persist, but peer-to-peer overlays, leveraging Distributed Hash Tables and event-based systems, offer potential solutions [7].

A distributed hash table (DHT) is a decentralized system that stores key-value pairs, enabling quick retrieval by any node. DHTs ensure scalability and resilience, distributing key-value mapping among nodes and handling node changes effectively, making them suitable for large-scale systems. Structured P2P networks using DHTs, like CAN [9], Chord [13], Pastry [11], and Tapestry [15], offer better load balancing, fault tolerance, and search efficiency compared to unstructured networks like Gnutella and Kaaza [8,10]. However, frequent node changes (churn) pose challenges. Hybrid systems combining structured and unstructured architectures [1,2,4,5,12] and non-DHT-based approaches, such as interest/resource-based systems [7], aim to manage churn while retaining DHT benefits.

In this context, the paper proposes a non-DHT fog computing architecture tailored for Healthcare 4.0, leveraging an interest/resource-based design for efficient resource sharing and sensor data processing. This architecture, inspired by previous work on interest/resource-based non-DHT systems [7], seeks to enhance performance and reduce the complexity associated with DHTs.

In a recent work [12], the authors propose a pioneering fog computing architecture tailored for efficient resource sharing within the Healthcare 4.0 framework. Departing from conventional DHT-based approaches, they advocate for a non-DHT RC-based, publish-subscribe mechanism to foster peer-to-peer (P2P) interactions. This architecture demonstrates superior data look-up latency compared to its DHT-based counterparts. However, a critical challenge arises from the centralization of the fog controller, posing a potential single point of failure in the network.

Addressing this concern, the paper delves into a novel location-based approach aimed at eliminating the reliance on the fog controller. By decentralizing control and leveraging location-aware mechanisms, the proposed solution aims to enhance network robustness and resilience. Moreover, the authors illustrate how this approach contributes to a reduction in the overall hop count for data lookup and transfer, thereby optimizing network efficiency and performance.

The significance of this contribution lies in its potential to mitigate the inherent vulnerabilities associated with centralized control points in fog computing architectures. By embracing a decentralized, location-based paradigm, the proposed approach not only enhances network reliability but also paves the way for more scalable and fault-tolerant Healthcare 4.0 systems.

2   RC Based P2P [7]

For illustration, let V symbolize a specific actor and R_i denote the resource category "movies." Therefore, $< R_i, V >$ represents movies (some or all) acted by a particular actor V.

Definition 1. *A resource is defined as a tuple $< R_i, V >$, where R_i stands for a resource's type and V for its value. Note that a resource might have several values.*

Definition 2. *In a peer-to-peer system with n different resource kinds, let S be the set of all peers. Then $S = \{C_i\}, 0 \le i \le n-1$, where C_i represents the subset of peers that have the same resource type R_i. This subset C_i is referred to as group i. We presume that C_i^h is the first peer among the peers in each group C_i to join the system, and the group-head of group C_i is referred to as C_i^h.*

The following two-level overlay architecture, as shown in Fig. 1, has been proposed in [7].

1. At level 1, there is a ring network consisting of the peers C_i ($0 \le i \le n-1$). The ring has n peers, corresponding to the number of different resource kinds. This ring network is known as a transit ring used for quick data search.
2. There are n fully connected networks (groups) of peers at level 2. Each group, say G_i, is made up of the peers of the subset C_{R_i} ($0 \le i \le n-1$), such that all peers in C_{R_i} are logically connected, and the network has a diameter of 1. Each G_i has a group-head C_i^h that connects it to the transit ring network.
3. Each peer on the transit ring network maintains a global resource table (GRT) with n tuples. The GRT contains one tuple per group, each of the form $<$Group-Head Logical address, IP address $>$.
4. Each group-head C_i^h maintains a local resource table (LRT) with k tuples, where k is the number of members in group G_i. The LRT contains tuples of the form $<$Group member Logical address, IP address$>$. This LRT is also maintained by all group members of G_i.
5. Any communication between a peer C_i in group G_x and C_j in group G_y occurs through the corresponding group-heads C_x^h and C_y^h.

2.1   Relevant Properties of Modular Arithmetic

Consider the set S_n of non-negative integers less than n, where $S_n = \{0, 1, 2, \ldots, n-1\}$. This is known as the residue set or residue classes

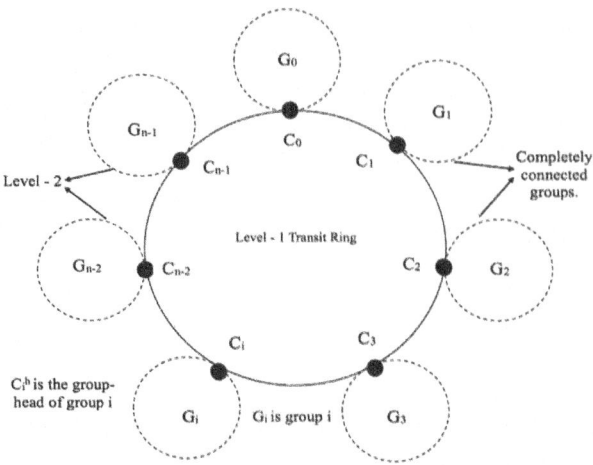

Fig. 1. A two-level RC-based structured P2P architecture with n distinct resource types.

(mod n). Each integer in S_n represents a residue class (RC). These residue classes are denoted by the symbols $[0], [1], [2], \ldots, [n-1]$, where $[r] = \{a : a$ is an integer, $a \equiv r \pmod{n}\}$.

For example, for $n = 3$, the classes are:

$$[0] = \{\ldots, -6, -3, 0, 3, 6, \ldots\}$$
$$[1] = \{\ldots, -5, -2, 1, 4, 7, \ldots\}$$
$$[2] = \{\ldots, -4, -1, 2, 5, 8, \ldots\}$$

Thus, any class $r \pmod{n}$ of S_n can be written as follows:

$$[r] = \{\ldots, (r - 2n), (r - n), r, (r + n), (r + 2n), \ldots\}$$

A few relevant properties of the residue class are stated below.

Lemma 31 *Any two numbers of any class r of S_n are mutually congruent.*

Proof. Consider any two numbers N' and N'' of class r. These numbers can be written as:

$$N' \equiv r \pmod{n}; \text{ therefore, } (N' - r)/n = \text{an integer, say } I' \tag{1}$$

$$N'' \equiv r \pmod{n}; \text{ therefore, } (N'' - r)/n = \text{an integer, say } I'' \tag{2}$$

Using (1) and (2), we obtain:

$$(N' - N'')/n = ((N' - r) - (N'' - r))/n = I' - I'' = \text{an integer}. \tag{3}$$

Therefore, $N' \equiv N'' \pmod{n}$;

Additionally, $N'' \equiv N' \pmod{n}$ because the congruence relation ($\equiv$) is symmetric. Hence, the proof.

2.2   Assignments of Overlay Addresses

Let us assume that there are n different types of resources in a resource/interest-based P2P system. Please note that n can be set to an extremely large value a priori to accommodate many distinct resource types. Consider the set of all peers in the system given as $S = \{C_{R_i}\}$ $(0 \leq i \leq n-1)$. Additionally, as mentioned earlier, for each subset C_{R_i} (i.e., group G_i) peer C_i is the first peer with resource type R_i to join the system.

In the suggested overlay architecture [7,12], the positive integers belonging to distinct classes are utilized to determine the following parameters:

1. Logical addresses of peers in a subset C_{R_i} (i.e., group G_i): It will be shown how to use these addresses to support the claim that all peers ($\in G_i$) are (logically) directly linked to one another, producing an overlay network of diameter 1. Each G_i, as used in graph theory, is a complete graph.
2. Identifying which peers on the transit ring network are neighbors with one another.
3. Identifying each distinct resource type with a unique code.

The assignment of logical addresses to the peers at the two levels and the resources happen as follows:

1. At level 1, the smallest non-negative number (r) of the residue class r (mod n) of the residue system, S_n, is assigned to each group-head C_r^h of group G_r.
2. At level 2, the group G_r (i.e., the subset C_{R_r}) will be formed by all peers with the same resource type R_r, with the group-head C_r^h connected to the transit ring network. Given to each new peer that joins group G_r is the group membership address $(r + j \cdot n)$, where j is $1, 2, 3, \ldots$.
3. Resource class R_r possessed by peers in G_r is assigned the code r, which is also the logical address of the group-head C_r^h of group G_r.
4. A corresponding tuple of <Group-Head Logical Address, IP Address> is added to the global resource table (GRT) each time a new group-head joins.

Remark 31 *GRT remains sorted with respect to the logical addresses of the group-heads.*

Definition 3. *Two peers C_i^h and C_j^h on the ring network are logically linked together if $(i+1) \mod n = j$.*

Remark 32 *The last group-head C_{n-1}^h and the first group-head C_0^h are neighbors based on Definition 3. It justifies that the transit network is a ring.*

Definition 4. *Two peers of a group G_r are logically linked together if their assigned logical addresses are mutually congruent.*

Remark 33 *The diameter of the transit ring network is $n/2$.*
Lemma 32 *Each group G_r forms a complete graph.*

Proof. A pair of peers in a group G_r are said to be logically connected by Definition 4 if their assigned logical addresses are compatible with one another. Additionally, from Lemma 1, we see that any two numbers of any class r of S_n are mutually congruent. Hence, every peer has direct logical connectivity with every other peer in the same group G_r. Thus, the evidence.

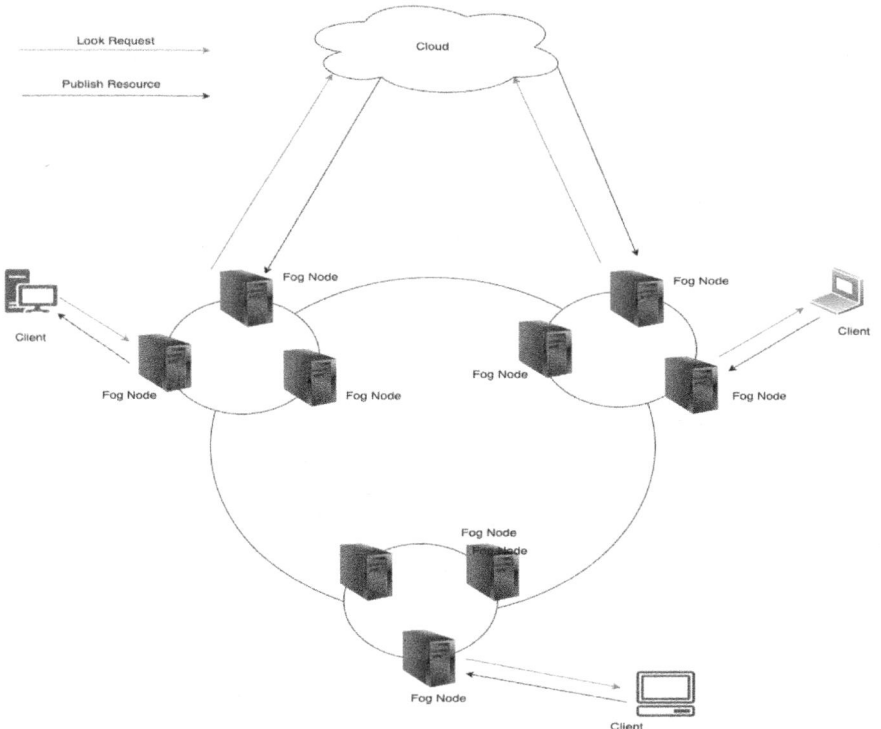

Fig. 2. RC-based structured P2P Fog Computing Architecture for Healthcare 4.0.

3   RC-Based Geographical P2P Fog Computing Architecture

The three-layer architecture of RC-based P2P fog computing consists of client nodes in the first layer, fog nodes in the second layer, and a cloud node in the third layer. Because fog nodes are now connected via P2P and utilise RC lookup techniques when searching for files, the RC-based P2P fog computing architecture employs an RC network topology to connect fog nodes and ease resource search and data transfers.

The architecture of RC-based peer-to-peer fog computing is depicted in Fig. 2. When a client node requests a resource or data from its associated fog node, which is determined based on geographical proximity, the architecture relies on Residue Class (RC) mechanisms for storing services. These RC mechanisms play a vital role in this architecture, ensuring efficient resource management and facilitating data retrieval. This approach leverages the concept of RC to optimize the allocation and distribution of resources within the fog computing network, taking advantage of its mathematical properties for effective data storage and retrieval.

In order to enable efficient resource discovery, we propose in this section to overlay two networking infrastructures on top of the actual fog nodes. In order to use the RC-based p2p architecture mentioned in Sect. 2, we have made the following modifications to our Healthcare 4.0 fog-computing architecture, which are detailed below.

The network engineer groups the heterogeneous network based on a *range of computing power*, using the concept from [12]. Lower logical addresses serve nodes with 1–2 cores and less than 256 MB of RAM, while higher addresses serve nodes with more capacity, such as 16–32 cores and several GB of RAM. The designer determines the logical address and number of groups according to application needs. This plan adapts to slight variations in group requirements, while significant changes prompt the node to notify its group head, who reassigns the node using the GRT.

To improve data lookup efficiency, a new column in the GRT represents the computing range for each group. In our RC-based fog-computing architecture, resource R_i corresponds to range i. For example, in a P2P fog computing network with three groups, the computing ranges are R_0, R_1, and R_2. Requests falling between R_1 and R_2 are sent to the group head with range R_2.

3.1   RC-Based Lookup Protocol

Assume an RC-based peer-to-peer system with n distinct computing ranges, with S being the set of all peers. Then $S = \{C_i\}, (0 \leq i \leq n - 1)$, where C_i denotes the subset of peers having computing capacity in the range R_i. In this study, this subset C_i is referred to as group i. Furthermore, we assume that C_i^h is the initial peer to join the system from each group C_i. The group-head of group C_i is known as C_i^h.

Assume that the IoT/client device (I_x) delivers some data request ReR_i to the nearest fog node F_i (where $F_i \in C_i$), depending on geographical proximity. At this time, four scenarios are possible:

- **Scenario 1:** The fog node $F_i \in C_i$ itself has ReR_i.
- **Scenario 2:** The fog node who has ReR_i is the group member of the group C_i.
- **Scenario 3:** The fog node who has ReR_i is the group-head of another group C_j^h.

- **Scenario 4:** The fog node who has ReR_i is the group member of another group C_j.

Below we present the algorithms for all the above scenarios.

- **Scenario 1:** The fog node $F_i \in C_i$ itself has ReR_i:
  - I_x found F_i is the closed fog node based on geographical proximity
  - I_x request ReR_i to F_i;
  - If ReR_i falls in the computing range R_i and the fog node F_i have the resource ReR_i it will provide service to I_x.

This scenario is presented in Algorithm 1.

Algorithm 1. Scenario 1: The fog node $F_i \in C_i$ itself has the resource

---
1: I_x found F_i is the closed fog node based on geographical proximity
2: I_x request ReR_i to F_i;
3: **if** $(ReR_i \in R_i) \wedge F_i$ possess ReR_i **then**
4:     F_i unicasts service to I_x;
5: **else if** Scenario 2 or Scenario 3 or Scenario 4 **then**
6:     Respective Scenario solution
7: **else**
8:     F_i forwards ReR_i to C_i^h;
9:     C_i^h forwards ReR_i to cloud node;
10:    cloud responds with ReR_i to I_x;
11: **end if**

Observation 1: The number of hops required for an I_x to find a resource in the proposed overlay P2P architecture for **Scenario 1** is only 1 which is constant.

- **Scenario 2:** The fog node who has ReR_i is the group member of the group C_i.
  - I_x found F_i is the closed fog node based on geographical proximity
  - I_x request ReR_i to F_i;
  - If ReR_i falls in the computing range R_i and the fog node F_i does not have the resource ReR_i it will broadcast the advertised message ReR_i and information of I_x in its group C_i using LRT.
  - Fog node A_i in C_i who has ReR_i will reply with the service to the I_x.

This scenario is presented in Algorithm 2.

Algorithm 2. Scenario 2: The fog node who has ReR_i is the group member of the group C_i

1: I_x found F_i is the closed fog node based on geographical proximity
2: I_x request ReR_i to F_i;
3: **if** $(ReR_i \in R_i) \wedge F_i$ does not possess ReR_i **then**
4:     F_i broadcast the advertised message tuple $\{ReR_i, I_x\}$ in its group C_i using LRT;
5:     **if** $A_i \in C_i$ possess ReR_i **then**
6:         A_i unicasts service to I_x;
7:     **else if** Scenario 3 or Scenario 4 **then**
8:         Respective Scenario solution
9:     **else**
10:         F_i forwards $\{ReR_i, I_x\}$ to cloud node;
11:         cloud responds with ReR_i to I_x;
12:     **end if**
13: **end if**

Observation 2: The number of hops required for an I_x to find a resource in the proposed overlay P2P architecture for **Scenario 2** is only 2, which is constant.

– **Scenario 3:** The fog node who has ReR_i is the group-head of another group C_j^h.
  • I_x found F_i is the closed fog node based on geographical proximity.
  • I_x request ReR_i to F_i.
  • Fog node F_i and any $A_i \in C_i$ does not have the resource ReR_i.
  • F_i forwards tuple $\{ReR_i, I_x\}$ to C_i^h
  • C_i^h determines the group-head C_y^h's address code from GRT such that $ReR_i \in$ the computing range R_y for C_y^h (i = y).
  • C_i^h computes $|i - j| = h$.
  • Based upon the value of h, it will forward ReR_i to its predecessor or its successor.
  • Every group-head traversed C_j^h forwards ReR_i until j = y.
  • If C_j^h has the resource, it will reply with the service to the I_x.

This scenario is presented in Algorithm 3.
Observation 3: The number of hops required for an FC to find a resource in the proposed overlay P2P architecture for **Scenario 3** is $(\frac{n}{2} + 2)$, where n is the total number of computing ranges in the network, the data-lookup complexity is $O(n)$.

– **Scenario 4:** The fog node who has ReR_i is the group member of another group C_j.
  • I_x found F_i is the closed fog node based on geographical proximity.
  • I_x request ReR_i to F_i.
  • Fog node F_i and any $A_i \in C_i$ does not have the resource ReR_i.
  • F_i forwards tuple $\{ReR_i, I_x\}$ to C_i^h

Algorithm 3. Scenario 3: The fog node who has ReR_i is the group-head of another group C_j^h

---

1: I_x found F_i is the closed fog node based on geographical proximity
2: I_x request ReR_i to F_i;
3: **if** $ReR_i \notin R_i$ **then**
4:     F_i unicasts tuple $\{ReR_i, I_x\}$ to C_i^h
5:     C_i^h determines the group-head C_y^h's address code from GRT;   ▷ $ReR_i \in$ the computing range R_y for C_y^h (i = y)
6:     C_i^h computes $|i - j| = h$;
7:     **if** h ¿ n/2 **then**   ▷ n = total no. of computing ranges
8:         C_i^h forwards the advertised message $\{ReR_i, I_x\}$ along with C_y^h's IP address to its predecessor C_{n-1}^h;
9:     **else**
10:        C_i^h forwards the advertised message $\{ReR_i, I_x\}$ along with C_y^h's IP address to its successor C_{i+1}^h;   ▷ Looking for minimum no. of hops along the transit ring network
11:    **end if**
12:    All intermediate group-heads C_j^h forwards until j = y ▷ no. of hops along the ring in the worst case is n / 2
13:    **if** C_j^h possess ReR_i **then**
14:        C_j^h unicasts service to I_x;
15:    **else if** Scenario 4 **then**
16:        Respective Scenario solution
17:    **else**
18:        C_i^h forwards $\{ReR_i, I_x\}$ to cloud node;
19:        cloud responds with ReR_i to FC;
20:    **end if**
21: **end if**

- C_i^h determines the group-head C_y^h's address code from GRT such that $ReR_i \in$ the computing range R_y for C_y^h (i = y).
- C_i^h computes $|i - j| = h$.
- Based upon the value of h, it will forward ReR_i to its predecessor or its successor.
- Every group-head traversed C_j^h forwards ReR_i until j = y.
- If C_j^h does not have ReR_i it will broadcast the message ReR_i in group C_y .
- Fog node A_y in C_y who has ReR_i will reply with the service to the I_x.

This scenario is presented in Algorithm 4.

Observation 4: The number of hops required for an FC to find a resource in the proposed overlay P2P architecture for **Scenario 4** is $(\frac{n}{2} + 3)$, where n is the total number of computing ranges in the network, the data-lookup complexity is $O(n)$.

Algorithm 4. Scenario 4: The fog node who has ReR_i is the group-head of another group C_j^h

---

1: I_x found F_i is the closed fog node based on geographical proximity
2: I_x request ReR_i to F_i;
3: **if** $ReR_i \notin R_i$ **then**
4:     F_i unicasts tuple $\{ReR_i, I_x\}$ to C_i^h
5:     C_i^h determines the group-head C_y^h's address code from GRT;   ▷ $ReR_i \in$ the computing range R_y for C_y^h (i = y)
6:     C_i^h computes $|i - j| = h$;
7:     **if** h ¿ n/2 **then**   ▷ n = total no. of computing ranges
8:         C_i^h forwards the advertised message $\{ReR_i, I_x\}$ along with C_y^h's IP address to its predecessor C_{n-1}^h;
9:     **else**
10:         C_i^h forwards the advertised message $\{ReR_i, I_x\}$ along with C_y^h's IP address to its successor C_{i+1}^h;   ▷ Looking for minimum no. of hops along the transit ring network
11:     **end if**
12:     All intermediate group-heads C_j^h forwards until j = y ▷ no. of hops along the ring in the worst case is n / 2
13:     **if** C_j^h does not possess ReR_i **then**
14:         C_j^h broadcast the advertised message $\{ReR_i, I_x\}$ in its group C_y using LRT;
15:         **if** $A_y \in C_y$ possess ReR_i **then**
16:             A_y unicasts service to I_x;
17:         **else**
18:             C_i^h forwards $\{ReR_i, I_x\}$ to cloud node;
19:             cloud responds with ReR_i to FC;
20:         **end if**
21:     **end if**
22: **end if**

---

4   Conclusion

Healthcare 4.0 shifts healthcare from manual procedures to technology-driven decision-making, prioritizing patient experience tech. IoT devices gather patient health data, initially in cloud-based centers, posing latency, privacy, and cost problems. Fog computing resolves these issues but faces challenges in resource location and utilization.

In this paper, we present geographical, resource/computing range-based, RC-based fog computing, a novel method for building a scalable Healthcare 4.0 fog-computing system that performs data-lookup operations with high efficiency, lowering latency and providing more rapid localized services. Because RC-based fog can perform more processing demands locally, fewer requests must be forwarded to the cloud, thereby saving Internet traffic.Compared to similar work presented in [12], our approach demonstrates that by eliminating the fog controller, we have reduced the number of hops required for clients to locate resources and avoided the single point of failure issue.Several cloud service types, including SaaS, PaaS, and IaaS, may be combined at various level 2 networks under the RC-based fog-computing paradigm. Furthermore, in our near future study, we want to simulate our suggested architecture utilizing fog simulators such as IFogSim2 [3] and create real-time data to demonstrate the system's efficacy in genuine Healthcare 4.0 situations.

References

1. Aekaterinidis, I., Triantafillou, P.: Pastrystrings: A comprehensive content-based publish/subscribe dht network. In: 26th IEEE International Conference on Distributed Computing Systems (ICDCS'06), pp. 23–23 (2006)
2. Gupta, A., Sahin, O.D., Agrawal, D., El Abbadi, A.: Meghdoot: content-based publish/subscribe over P2P networks. In: Jacobsen, H.-A. (ed.) Middleware 2004. LNCS, vol. 3231, pp. 254–273. Springer, Heidelberg (2004). https://doi.org/10.1007/978-3-540-30229-2_14
3. Gupta, H., Vahid Dastjerdi, A., Ghosh, S.K., Buyya, R.: ifogsim: A toolkit for modeling and simulation of resource management techniques in internet of things, edge and fog computing environments (2016)
4. Gupta, H., Vahid Dastjerdi, A., Ghosh, S.K., Buyya, R.: ifogsim: A toolkit for modeling and simulation of resource management techniques in the internet of things, edge and fog computing environments. Software: Practice and Experience **47**, 1275 – 1296 (2016)
5. Hao, Z., Novak, E., Yi, S., Li, Q.:. Challenges and software architecture for fog computing. IEEE Internet Computing **21**(2), 44–53 (2017)
6. Hathaliya, J.J., Tanwar, S., Tyagi, S., Kumar, N.:. Securing electronics healthcare records in healthcare 4.0 : A biometric-based approach. Comput. Electr. Eng. **76**, 398–410 (2019)
7. Kaluvakuri, S., Gupta, B., Rekabdar, B., Maddali, K., Debnath, N.: Design of RC-based low diameter two-level hierarchical structured p2p network architecture. In: Serrhini, M., Silva, C., Aljahdali, S. (eds.) Innovation in Information Systems and Technologies to Support Learning Research: Proceedings of EMENA-ISTL 2019, pp. 312–320. Springer International Publishing, Cham (2020). https://doi.org/10.1007/978-3-030-36778-7_34
8. Liang, J., Kumar, R., Ross, K.W.: The kazaa overlay : A measurement study (2004)
9. Ratnasamy, S., Francis, P., Handley, M., Karp, R., Shenker, S.: A scalable content-addressable network. In: Proceedings of the 2001 Conference on Applications, Technologies, Architectures, and Protocols for Computer Communications, SIGCOMM '01, pp. 161–172, New York, NY, USA, 2001. Association for Computing Machinery
10. Ripeanu, M.: Peer-to-peer architecture case study: Gnutella network. In: Proceedings First International Conference on Peer-to-Peer Computing, pp. 99–100 (2001)
11. Rowstron, A., Druschel, P.: Pastry: Scalable, decentralized object location and routing for large-scale peer-to-peer systems, vol. 2218, pp. 329–350, 01 (2001)
12. Roy, I., Mitra, R., Rahimi, N., Gupta, B.L Efficient non-dht-based rc-based architecture for fog computing in healthcare 4.0. IoT **4**(2), 131–149 (2023)
13. Stoica, I., Morris, R., Karger, D., Kaashoek, M.F., Balakrishnan, H.: Chord: A scalable peer-to-peer lookup service for internet applications. In: Proceedings of the 2001 Conference on Applications, Technologies, Architectures, and Protocols for Computer Communications, SIGCOMM '01, pp. 149–160, New York, NY, USA (2001). Association for Computing Machinery

14. Vora, J., et al.: Ensuring privacy and security in e- health records. In: 2018 International Conference on Computer, Information and Telecommunication Systems (CITS), pp. 1–5 (2018)
15. Zhao, B.Y., Kubiatowicz, J.D., Joseph, A.D.: Tapestry: An Infrastructure For Fault-tolerant Wide-area Location and Technical report, USA (2001)

Applying Skyline Operator and Taxicab Geometry to Identify Optimal Locations for Establishing Business Properties

Leena Jana Ghosh[1] (ID), Takaaki Goto[2], Subhankar Roy[3] (ID), Subhashis Das[3] (ID), Mainak Sen[4], and Partha Ghosh[3]([✉]) (ID)

[1] JC Edutech, Naihati, West Bengal, India
[2] Toyo University, Saitama, Japan
tg@gotolab.net
[3] Academy of Technology, Adisaptagram, West Bengal, India
pghosh44@gmail.com
[4] Techno India University, Kolkata, West Bengal, India

Abstract. A proper multi-criteria optimization for any business organization always improvise business intelligence. Handling multiple criteria of customers' through any structured query language is difficult, especially when the criterion are inversely proportional to each other. Handling multi-criteria optimization through skyline operator may be one of the possible best solutions for handling criterion those are inversely proportional. Generally, skyline operator determines an optimal point in the problem domain and ranks other skyline points by computing the shortest distance from that optimal point. But this type of approach is not suitable for applying the skyline operator in the domain of travelling and tourism industry as there may be several physical obstacles upon the shortest distance path. This paper proposes a novel methodology for establishing new business properties or hotels that satisfy all travellers' and property owners' requirements, allowing travellers to visit all their desired locations.

Keywords: Multi-criteria optimization · Skyline operator · Business intelligence · Business property. Tour · Travel

1 Introduction

In the dynamic landscape of business intelligence, optimizing multiple criteria is essential for enhancing organizational performance. Businesses strive to meet diverse customer requirements [12], which often present conflicting demands. For instance, the general goal for the travellers is to pay minimum amount for the hotel cost, while trying to stay closure to the visiting points with all premium facilities [1]. Traditional methods using structured query language (SQL) face significant challenges in effectively addressing these multi-criteria optimizations. One promising approach to overcoming these challenges is the application of the skyline operator. The skyline operator [1, 3, 5, 15] is a powerful tool in multi-criteria optimization. It works by identifying a set of

G. Hu et al. (Eds.): CAINE 2024, CCIS 2242, pp. 202–209, 2025.
https://doi.org/10.1007/978-3-031-76273-4_16

optimal points, known as skyline points, which are not dominated by any other points in the dataset based on the specified criteria. These points represent the most preferred solutions from a multi-criteria perspective. The operator then ranks these points by calculating the shortest distance from an optimal reference point, thereby aiding in the decision-making process.

However, this conventional approach has limitations when applied to specific industries, such as the travel and tourism sector. In the travel and tourism industry, multiple criteria optimization involves unique complexities. Travellers have varied preferences and requirements that must be met to ensure customer satisfaction. These requirements might include proximity to tourist attractions, availability of amenities, cost considerations, and more. Unlike in some other sectors, the physical layout and geographical constraints significantly impact the optimization process. For instance with references to Fig. 1, obstacles such as rivers, mountains, or urban infrastructure can hinder direct paths, making the shortest distance calculation less practical and sometimes infeasible.

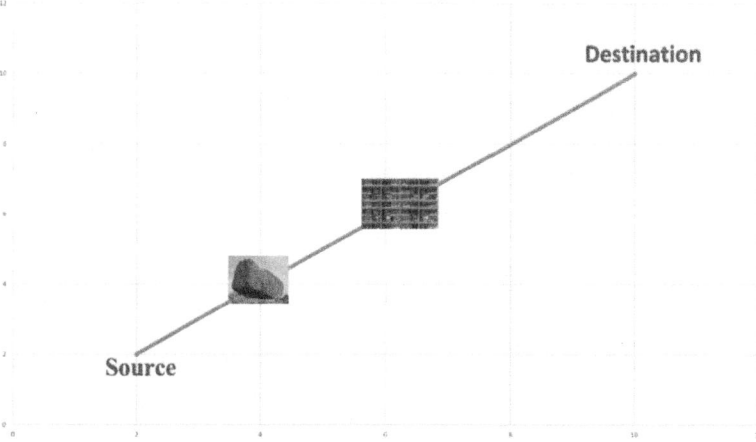

Fig. 1. Obstacles upon shortest distance path

Considering these challenges, there is a need for a more sophisticated methodology that can effectively handle multi-criteria optimization in the travel and tourism domain. In this regard, taxicab geometry [7, 10] based distance calculation can resolve this problem that calculate distance between two points is the sum of the absolute differences of their coordinates. To calculate the taxicab distance between two points (X_1, Y_1) and (X_2, Y_2), formula is shown in equation-1.

$$D_{taxicab} = |X_1 - Y_1| + |X_2 - Y_2| \tag{1}$$

Therefore, the taxicab path doesn't follow the direct 'as-the-bird-flies' path but rather the path along the grid lines. Hence, if there are any physical obstacles, it can find an alternative taxicab path. Most importantly, all taxicab paths have the same distance [7, 17].

This paper proposes a novel approach aimed at establishing new business properties, such as hotels, that cater to all travellers' needs while considering geographical and infrastructural constraints. The proposed methodology leverages the skyline operator in a manner that adapts to the unique demands of the travel and tourism industry. The methodology integrates Taxicab Distance [7, 10] to account for physical obstacles and real-world distances. By incorporating spatial data and advanced computational techniques, the approach ensures that the optimized [18] solutions are practical and feasible. The ultimate goal is to identify optimal locations for new hotels that maximize traveller satisfaction by providing convenient access to desired tourist destinations and essential services. This novel methodology addresses the shortcomings of traditional skyline approaches in handling multi-criteria optimization within the constraints of physical geography. This research aims to pave the way for more efficient and customer-centric [12] business property establishment in the travel and tourism industry.

This paper is organized in the following way: the related study is discussed in Sect. 2, followed by the proposed methodology in Sect. 3. We perform a case study on a real-world dataset to check the efficiency of the proposed model. Finally, in Sect. 5, we conclude.

2 Related Work

Borzsonyi et al. [1] first proposed Skyline computation [3–5, 14] using the D&C algorithm [1], dividing datasets into smaller sections, and the BNL algorithm [1], which uses a window to compare and eliminate dominated tuples. Initially, skyline computation began by optimizing two dimensions relative to a single point [1, 5, 11]. The concept of the Best Skyline Point (BSP) [2] was subsequently introduced. Once the BSP is identified, other points are ranked based on their proximity to it, minimizing coordinate dissimilarities. This [2] dynamic method performs computations during query execution and is suitable for uncertain databases [2, 3]. A framework [8] for processing subspace skyline queries on streaming data with a validity time window was further proposed. This approach [8] uses an index structure that updates at regular intervals. Queries are evaluated based on the latest indexed data, making results approximate but consistent with standard streaming data practices. Skyline queries over crowd-sourced incomplete databases [16] are addressed with an approach for computation and missing values estimation. It [16] employs data filtration and an AFD-based strategy, minimizing costs and maintaining accuracy while reducing the need for crowd-based estimation. A novel approach [9] efficiently addresses challenges like data stragglers, skew, and high computation costs in distributed skyline query processing. Employing a three-phase method [9], it partitions data along the Z-order curve, minimizes intermediate data, computes skyline candidates in parallel, and merges them, achieving over tenfold performance improvement in extensive experiments. The main problem of increasing dimensions for skyline computation is the time complexity. Hence, a novel methodology [6] is proposed to compute Skyline both for multiple points and multiple dimensions. This framework [6] uniquely applies the skyline method to multiple dimensions without exceeding $O(n^3)$ complexity. It's mainly used in travel tourism to optimize hotel searches based on various factors like distance, location, taxi fare, room fare, and other costs. It can find multiple hotels even when points of interest are many and spread out. An enhanced dynamic skyline query method for road networks [13] is introduced, utilizing a GD-tree index for efficient location storage, optimal scan endpoints for reduced computation, and histogram-based statistics. Extensive experiments confirm its [13] efficiency on real road network datasets. But, when applying skyline computation in the Euclidean plane [7, 10], encountering physical obstacles along the shortest distance path poses a challenge. An alternative solution is the Taxicab path [7, 10], which calculates the shortest path using Manhattan distance. This ensures that if any physical obstacles obstruct the shortest path, alternative Taxicab paths exist, with the crucial feature that all Taxicab paths have the same distance. This uniformity simplifies decision-making and ensures consistent results regardless of the specific path chosen. Thus, Taxicab paths offer a reliable workaround for navigating around obstacles, making skyline computation more robust and applicable in real-world scenarios with complex environments.

3   Proposed Model

This proposed model starts by collecting coordinates of visiting points and available vacant spaces on a two-dimensional plane (longitude and latitude) for establishing new business properties. Thereafter, it gathers user/promoter requirements such as the types of properties to be established, the area of vacant space required, the permissible taxicab distance from all visiting points, etc. Subsequently, it categorizes requirements into three dimensions, namely the area of vacant spaces, the cumulative taxicab distance of all visiting points from the vacant spaces, and other user/promoter requirements that are user-specific. It then applies skyline computation to retain those vacant spaces which are non-dominated by any other point within that three-dimensional space. Finally, it ranks all skyline points (vacant spaces) with respect to the median of the three-dimensional space.

Algorithm: Business Property Location Optimization

Input:
1.   Coordinates of visiting points (VP_i).
2.   Coordinates of available vacant spaces (VS_j).
3.   Other user/promoter requirements.

Output:
   Ranked list of optimal vacant spaces.

Steps:
1.   /* **Collect Coordinates** */
   Input the coordinates of all visiting points ($V P_i$) and all available vacant spaces ($V S_j$).
2.   /* **Gather User/Promoter Requirements** */
   2.1 Input the type of property to be established.
   2.2 Input the area of vacant space required.
   2.3 Input the permissible taxicab distance from all visiting points.

3. /* **Categorize Requirements into Three Dimensions** */

3.1 Dimension 1: Area of vacant spaces A_j.

3.2 Dimension 2: Cumulative taxicab distance of all visiting points from vacant spaces D_j

3.2.1 For each VS_j, calculate the cumulative taxicab distance:

$$D_j = \sum |VS_j.x - VP_i.x| + |VS_j.y - VP_i.y|$$

3.3 Dimension 3: Other user/promoter requirements R_j.

4 /* **Apply Skyline Computation** */

4.1 For each vacant space VS_j:

4.1.1 Compare VS_j with all other vacant spaces VS_k.

4.1.2 VS_j is dominated by VS_k iff.
$(A_k \leq A_j)$ and $(D_k \leq D_j)$ and $(R_k \leq R_j)$,

with at least one strict inequality.

4.2 Retain VS_j iff. it is not dominated by any other VS_k.

5 /* **Rank Skyline Points** */

5.1 Calculate the median values of the three dimensional plane.

5.2 Rank the non-dominated vacant spaces (skyline points) based on their proximity to the median values in the three-dimensional plane.

6 /* **Output the Ranked List** */

Provide the ranked list of optimal vacant spaces.

4 Case Study

In this section, we have applied our proposed model upon a real world dataset. Experiments were conducted on a Windows machine with an Intel Core i5 processor, 16 GB of RAM, and a 250 GB SSD. The experimental setup included various software tools, such as Jupyter Notebook and Python 3.11.3 for coding. For data manipulation, the pandas and numpy libraries were used, while matplotlib was utilized for data visualization. We collected information on vacant land in Edmonton, the capital city of Alberta, Canada. The data was sourced from the "Open Data Network" a reputable provider of data for researchers. The "Vacant Land Inventory data" is available at: https://www.opendatan etwork.com/dataset/data.edmonton.ca/svsw-2ub7. Further, the tourist spots of Edmonton city are collected from "City of Edmonton's Open Data Portal" (Data link: https:// data.edmonton.ca/Facilities-and-Structures/Attractions/7yt8-7467/data_preview).

This case study started with plotting of available vacant spaces for new property establishment in a two dimensional plane namely "Size of available vacant space" and "Taxicab distance in KM" as shown in Fig. 2.

Here the longitudinal and latitudinal distance are converted in Kilometre through the tools provided by National Hurricane Center (Link: https://www.nhc.noaa.gov/gccalc. shtml).

Next, by applying the proposed algorithm, we got the non-dominated points (available vacant spaces) as shown in the Fig. 3. In this case, for hotel establishment, the third

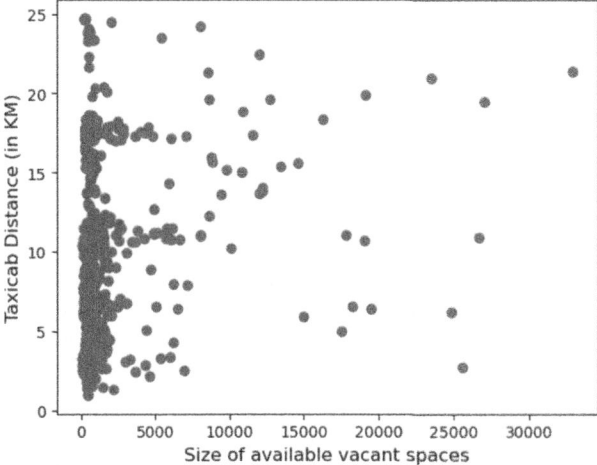

Fig. 2. Available vacant space plotted in a two-dimensional place Taxicab-distance & Size

dimensional constraints are considered as available vacant space in between 300 m² to 500 m² and OWNERSHIP_TYPE must be PRIVATE.

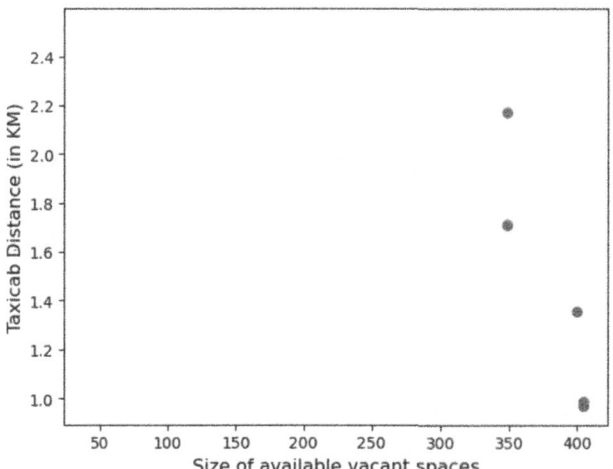

Fig. 3. Non-dominating points after applying Skyline computation

Therefore, the final output displayed to the user must be as shown in Fig. 4.

`Out[35]:`

	ADDRESS	NEIGHBOURHOOD_NAME	SIZE (m2)	OWNERSHIP_TYPE	LATITUDE	LONGITUDE
0	11008 86 AVENUE NW	Garneau	399.746	PRIVATE	53.522215	-113.515456
1	10251 114 STREET NW	Oliver	348.226	PRIVATE	53.544191	-113.517093
2	11120 86 AVENUE NW	Garneau	404.503	PRIVATE	53.522220	-113.518794
3	10137 115 STREET NW	Oliver	348.350	PRIVATE	53.541884	-113.518934
4	10137 115 STREET NW	Oliver	348.369	PRIVATE	53.541884	-113.518934

Fig. 4. Final output of the case study

5 Conclusion

This paper proposed a novel methodology utilizing a skyline operator for optimizing multi-criteria decision-making in the domain of the travel and tourism industry. The proposed model addresses the challenge of inversely proportional criteria by collecting coordinates of potential visiting points and available vacant spaces for establishing new business properties, along with user/promoter requirements. By categorizing these requirements into three dimensions namely: vacant space area, cumulative taxicab distance, and other user-specific needs, the model applies skyline computation to identify non-dominated vacant spaces. The methodology effectively simplifies the complexity of multi-criteria optimization by focusing on non-dominated points, ensuring that the selected spaces meet all the critical requirements without being surpassed by other options. Ranking these points with respect to the median of the three-dimensional space ensures a balanced consideration of all factors. Consequently, this approach provides a robust solution for establishing new business properties/hotels that satisfy travellers' diverse needs and facilitate convenient access to desired locations.

Experimental results on real-world data indicate that this optimized model significantly enhances business intelligence in the travel and tourism industry by facilitating more informed and strategic decision-making.

References

1. Borzsonyi, S., Kossmann, D., Stocker, K.: The skyline operator. In: International Conference on Data Engineering (ICDE) (2001)
2. Ghosh, P., Sen, S.: Ranking skyline points by computing nearest neighbor of best skyline point. In: IEEE India International Conference (INDICON) (2015)
3. Liu, X., Yang, D.-N., Ye, M., Lee, W.-C.: U-Skyline: a new skyline query for uncertain databases. In: IEEE Transactions on Knowledge and Data Engineering (2013)

4. Ghosh, P., Sen, S.: An alternative solution to skyline operation to reduce computational complexity. In: Second IEEE International Conference on Research in Computational Intelligence and Communication Networks (2016)

5. Van Craenendonck, T., Blockeel, H.: Constraint-based clustering selection. Mach. Learn. **106**, 1497–1521 (2017)

6. Ghosh, P., Sen, S., Cortesi, A.: Skyline computation over multiple points and dimensions. Innovat. Syst. Softw. Eng. **17**, 141–156 (2021). https://doi.org/10.1007/s11334-020-00376-1

7. Ghosh, P., Goto, T., Sen, S.: Taxicab geometry based analysis on skyline for business intelligence. Int. J. Softw. Innov. **6**(4) (2018)

8. Alami, K., Maabout, S.: A framework for multidimensional skyline queries over streaming data. Data Knowl. Eng. **127** (2020)

9. Tang, M., Yu, Y., Aref, W.G., Malluhi, Q.M., Ouzzani, M.: Efficient parallel skyline query processing for high-dimensional data. IEEE Trans. Knowl. Data Eng. **30**(10), 1838–1851 (2018)

10. Ghosh, P., Goto, T., Ghosh, L.J., Maji, G., Sen, S.: A novel approach to organize blood donation camp and blood unit wastage management. Int. J. Softw. Innovat. **12**(1) (2024). https://doi.org/10.4018/IJSI.333517

11. Rudenko L., Endres M.: Real-time skyline computation on data streams. In: Benczúr, A., et al. (eds.) New Trends in Databases and Information Systems. CCIS, vol. 909. Springer, Cham (2018).

12. Ghosh, P., Samanta, O., Goto, T., Sen, S.: Sales forecasting of overrated products: fine tuning of customer's rating by integrating sentiment analysis. IEEE Access **12**, 69578–69592 (2024). https://doi.org/10.1109/ACCESS.2024.3402133

13. Bai, M., Wang, Q., Chang, S., et al.: Location-based skyline query processing technology in road networks. J. Supercomput. 3183–3321 (2024). https://doi.org/10.1007/s11227-023-05563-y

14. Amiruzzaman, M., Jamonnak, S.: Multi-dimensional skyline query to find best shopping mall for customers. In: Conference on Data Science and Machine Learning Applications (CDMA), pp. 71–76 (2020)

15. Yuan, D., Zhang, L., Li, S., et al.: Skyline query under multidimensional incomplete data based on classification tree. J. Big Data (2024). https://doi.org/10.1186/s40537-024-00923-8

16. Swidan, M.B., Alwan, A.A., Turaev, S., Ibrahim, H., Abualkishik, A.Z., Gulzar, Y.: Skyline queries computation on crowdsourced-enabled incomplete database. IEEE Access **8**, 106660–106689 (2020). https://doi.org/10.1109/ACCESS.2020.3000664

17. Ghosh, P., Goto, T., Sen, S.: Computing skyline using taxicab geometry. Appl. Comput. Inf. Technol. 7–12 (2017). https://doi.org/10.1109/ACIT-CSII-BCD.2017.35

18. Roy, S., Mukhopadhyay, A.: A randomized optimal k-mer indexing approach for efficient parallel genome sequence compression. Gene **907** (2024)

Cryptography and Pseudorandom Generators

A New Pseudorandom Binary Generator Based on Nonlinear Feedback Registers with Boolean Function Combined Using Multiplexers

Narayan Debnath[1], Andrés Francisco Farías[2(✉)], Andrés Alejandro Farías[2],
Ana Gabriela Garis[3], Daniel Riesco[3], and Germán Antonio Montejano[3]

[1] School of Computing and Information Technology, Easten International University,
Thu Dau Mot, Vietnam
narayan.debnath@eiu.edu.vn
[2] Academic Department of Physical, Mathematical and Natural Sciences, National University
of La Rioja, La Rioja, Argentina
afarias665@yahoo.com.ar
[3] Department of Computer Science, Faculty of Physical-Mathematical and Natural Sciences,
National University of San Luis, San Luis, Argentina
{agaris,driesco,gmonte}@unsl.edu.ar

Abstract. Binary generators are devices responsible for delivering random or pseudo-random binary sequences, very useful in different branches such as cryptography, simulation, mathematics. These random chains must have high periods and high linear complexity and must also pass statistical tests to ensure that they are effectively random. The generator design process must take the above into account and must be verified at each stage for the final result to be successful. Sometimes the simple combination of poorly designed cryptographic components can lead to a faulty generator. This work shows the steps to follow for the successful development of a good generator. For the generator proposed in this presentation, eight nonlinear feedback shift registers (NLFSR) were used, which are cryptographic components with nonlinear Boolean filtering functions, which were combined by using five multiplexers. Finally, with the device obtained and operating with different keys, pseudorandom binary sequences were obtained that passed the randomness tests to which they were subjected.

Keywords: NLFSR. Multiplexer. Key. Boolean function · Statistical tests

1 Introduction

The work consists of the development of a pseudorandom binary generator based on the combination of Non Linear Feedback Shift Register (NLFSR) [1, 2], of different lengths, with their respective connection primitive polynomials, by using four-input multiplexers.

The procedure for developing a pseudo-random binary generator of these characteristics consists of several stages, which are indicated below:

G. Hu et al. (Eds.): CAINE 2024, CCIS 2242, pp. 213–227, 2025.
https://doi.org/10.1007/978-3-031-76273-4_17

- Design of the generator

  - Definition of the generator

- Components that make up the generator

  - Characteristics of the NLFSR
  - Characteristics of the four-input multiplexer and two control lines
  - Choice of the LFSR
  - Selection of four-variable Boolean functions based on their optimal cryptographic properties

- Key

  - Procedure for generating the initial states of the NLFSR
  - Permutation

- Composition of the generator with the elements already selected
- Statistical tests of randomness

  - Choice of the statistical tests of randomness to be used
  - Sequences, significance level and null hypothesis
  - Criteria for analyzing the results

- Performing the tests

  - Proportion of samples that pass tests
  - Statistical Tests (1 to 14) for Random and Pseudorandom Number
  - Random Excursions Test
  - Random Excursions Variant Test

2 Generator Design

2.1 Definition of the Generator

The generator proposed in this work is composed of eight NLFSR with connection primitive polynomials, and nonlinear Boolean filtering functions, whose binary outputs are combined with four four-input multiplexers, which in turn deliver four binary strings that are combined with a four-input multiplexer, resulting in a final pseudorandom binary sequence, as shown in Fig. 1.

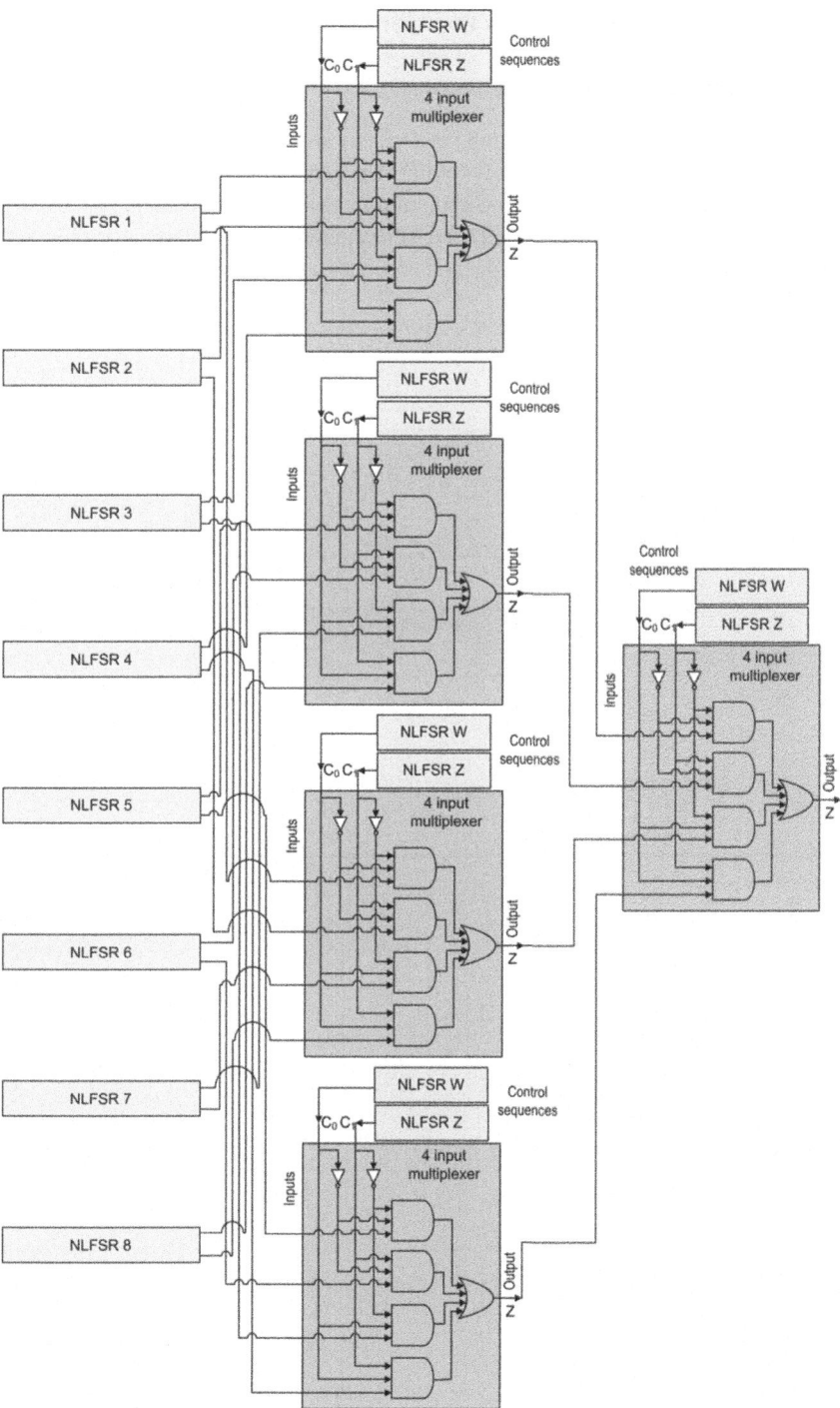

Fig. 1. Pseudorandom binary generator scheme

3   Components that make up the Generator

3.1   Characteristics of the NLFSR

The adopted NLFSR has the following structure as shown in Fig. 2, starting from an LFSR with a coupled connecting polynomial that generates the linear feedback. The polynomial is primitive, to achieve the maximum period of the sequence.

The feedback sequence is subjected to an XOR operation with the sequence generated by the four-variable boolean function, which produces a non-linear filtering, fed by some of the LFSR registers.

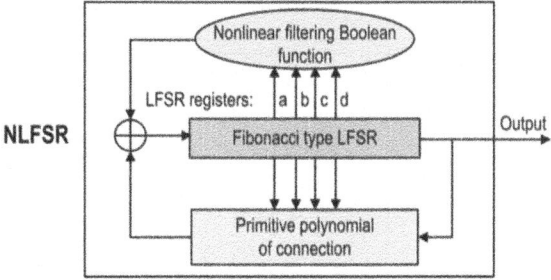

Fig. 2. NLFSR Scheme

3.2   Characteristics of the Four-Input Two-Control Line Multiplexer

The multiplexer has four inputs with two control sequences [3], Fig. 3:

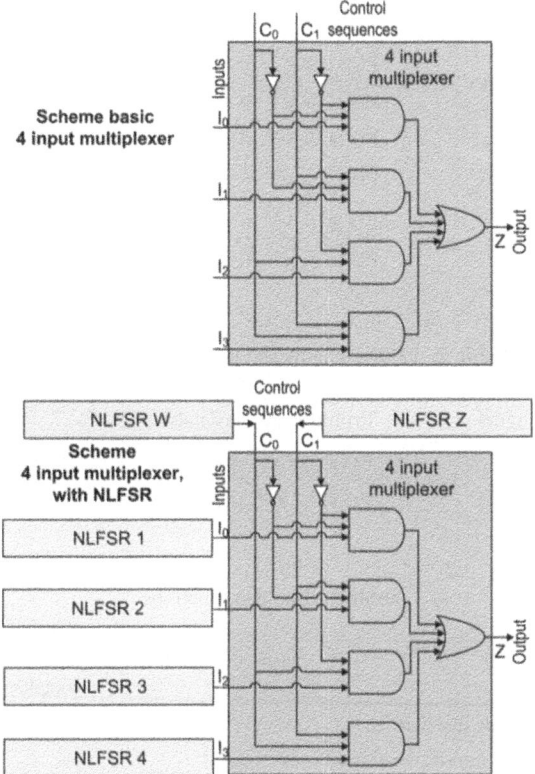

Fig. 3. Multiplexer Scheme

3.3   Choice of the LFSR

The lengths and primitive polynomials [4, 5] of the LFSR used to feed the multiplexers and to control them are indicated in Table 1 and Table 2:

3.4   Boolean Function Selection

The selected Boolean functions are shown in Table 3:

Desirable Cryptographic Properties. A Below are some of the most cryptographically significant properties, adopted for this work [6–8]:

- Balanced Function

Table 1. LFSR, lengths and primitive polynomials

LFSR	Lengths	Primitive polynomials
1	47	$P(x)_1 = x^{47} + x^{42} + x^{32} + x^{19} + x^{17} + x^5 + 1$
2	41	$P(x)_2 = x^{41} + x^{40} + x^{32} + x^{20} + x^{12} + x^{11} + 1$
3	31	$P(x)_3 = x^{31} + x^{16} + x^{14} + x^{10} + x^8 + x^1 + 1$
4	29	$P(x)_4 = x^{29} + x^{22} + x^{16} + x^{15} + x^{11} + x^3 + 1$
5	53	$P(x)_5 = x^{53} + x^{51} + x^{37} + x^{29} + x^{18} + x^4 + 1$
6	43	$P(x)_6 = x^{43} + x^{35} + x^{32} + x^{30} + x^{25} + x^8 + 1$
7	37	$P(x)_7 = x^{37} + x^{33} + x^{31} + x^{30} + x^{21} + x^3 + 1$
8	23	$P(x)_8 = x^{23} + x^{17} + x^{13} + x^{12} + x^{11} + x^5 + 1$

Table 2. LFSR, lengths and primitive polynomials

LFSR	Lengths	Primitive polynomials
W	17	$P(x)_W = x^{17} + x^{13} + x^8 + x^6 + x^5 + x^2 + 1$
Z	19	$P(x)_Z = x^{19} + x^{18} + x^{16} + x^{13} + x^{12} + x^6 + 1$

Table 3. Boolean functions

NLFSR	f_{NAF}	Balanced	Non-linearity	SAC compliant	a	b	c	d
	Nonlinear filtering functions				Registers			
1	$f_{84} = a \bullet c \oplus b \bullet c \oplus a \bullet d \oplus b \bullet d \oplus c \bullet d$	yes	4	yes	6	16	25	31
2	$f_{89} = a \bullet c \oplus b \bullet c \oplus d \oplus a \bullet d \oplus b \bullet d$	yes	4	yes	1	17	23	35
3	$f_{100} = a \bullet c \oplus b \bullet c \oplus d \oplus a \bullet b \bullet d \oplus c \bullet d$	yes	4	yes	1	6	12	28
4	$f_{176} = a \bullet c \oplus b \bullet c \oplus d \oplus a \bullet d \oplus b \bullet d$	yes	4	yes	1	7	13	27
5	$f_{199} = c \oplus a \bullet c \oplus b \bullet c \oplus d \oplus c \bullet d$	yes	4	yes	3	10	16	26
6	$f_{381} = c \oplus a \bullet b \bullet c \oplus a \bullet d \oplus b \bullet d \oplus c \bullet d$	yes	4	yes	6	14	27	34
7	$f_{468} = c \oplus d \oplus a \bullet d \oplus b \bullet d \oplus c \bullet d$	yes	4	yes	3	9	18	25
8	$f_{536} = a \bullet b \oplus b \bullet c \oplus a \bullet d \oplus b \bullet d \oplus c \bullet d$	yes	4	yes	3	17	19	20
W	$f_{541} = a \bullet b \oplus b \bullet c \oplus d \oplus a \bullet d \oplus c \bullet d$	yes	4	yes	3	9	12	15
Z	$f_{547} = a \bullet b \oplus b \bullet c \oplus d \oplus b \bullet d \oplus a \bullet c \bullet d$	yes	4	yes	3	7	15	16

- High non-linearity
- Meets strict avalanche criteria (SAC)

For the selection, the desirable cryptographic properties indicated in the previous paragraph are taken into account. Boolean functions of four variables are adopted.

4 Key

4.1 Procedure for Generating the Initial States of the NLFSR

To establish the initial states of the different NLFSR, a process is carried out that uses a 32-character key, which, expressed in ASCII code (American Standard Code for Information Interchange), has a length of 256 bits. The cryptographic procedure is presented in Fig. 4.

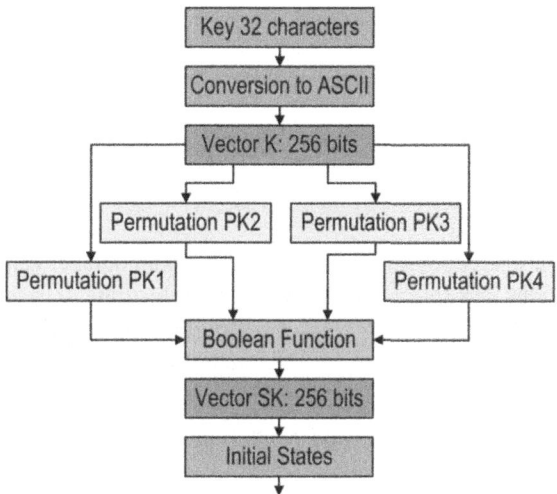

Fig. 4. Key to the generator

The operation results in a 256-bit vector SK[j], which will provide the initial states of the LFSR, sequentially.

4.2 Permutation

The permutations are calculated with a multiplicative congruent generator [9]. The generator has the following expression:

$$x_{i+1} = (a_x \cdot x_i) mod\, m_x$$

$$a_x = multiplier\ m_x = module\ x_0 = seed$$

Table 4 shows the values of the vectors, modules, multipliers and seeds:

5 Composition of the Generator with the Elements Already Selected

With the previously selected components, the structure of the pseudorandom binary generator is completed, Fig. 5.

Table 4. Vectors, modules, multipliers and seeds

Vector	module	multiplier	seed
PK1	1048576	1747	3249
PK2	1048576	1753	3271
PK3	1048576	1759	3301
PK4	1048576	1777	3347

6 Statistical Tests of Randomness

6.1 Choice of the Statistical Tests of Randomness to Be Used

Table 5 shows the statistical tests for random and pseudorandom numbers that make up this standard.

The statistical test suite for random and pseudorandom number generators for cryptographic applications was selected from the National Institute of Standards and Technology (NIST) Special Publication 800–22 Revision 1a, from the work of Rukhin (et al.) [10].

6.2 Sequences, Significance Level and Null Hypothesis

A total of one hundred binary sequences of 1,000,000 bits each, obtained from the binary generator from one hundred different keys, were subjected to the test set.

The significance level selected for all statistical tests is: $\alpha = 0.01$

The null hypothesis is: $H_0 \rightarrow p\_value > 0.01$

6.3 Criteria for Analyzing the Results

For the interpretation of the results, NIST 800–22 establishes two processes for this purpose, they are:

- Proportion of samples that pass tests
- Test for Uniformity of p-value

We adopted, for the interpretation of the test results, the first procedure: proportion of samples passing the tests.

7 Performing the Tests

7.1 Proportion of Samples that Pass Tests

To evaluate the results, the proportion of samples that pass the different tests is calculated, and a dot plot is made with this data. The plotted points must be located within the upper and lower limits; if this is the case, it means that the tests were successful.

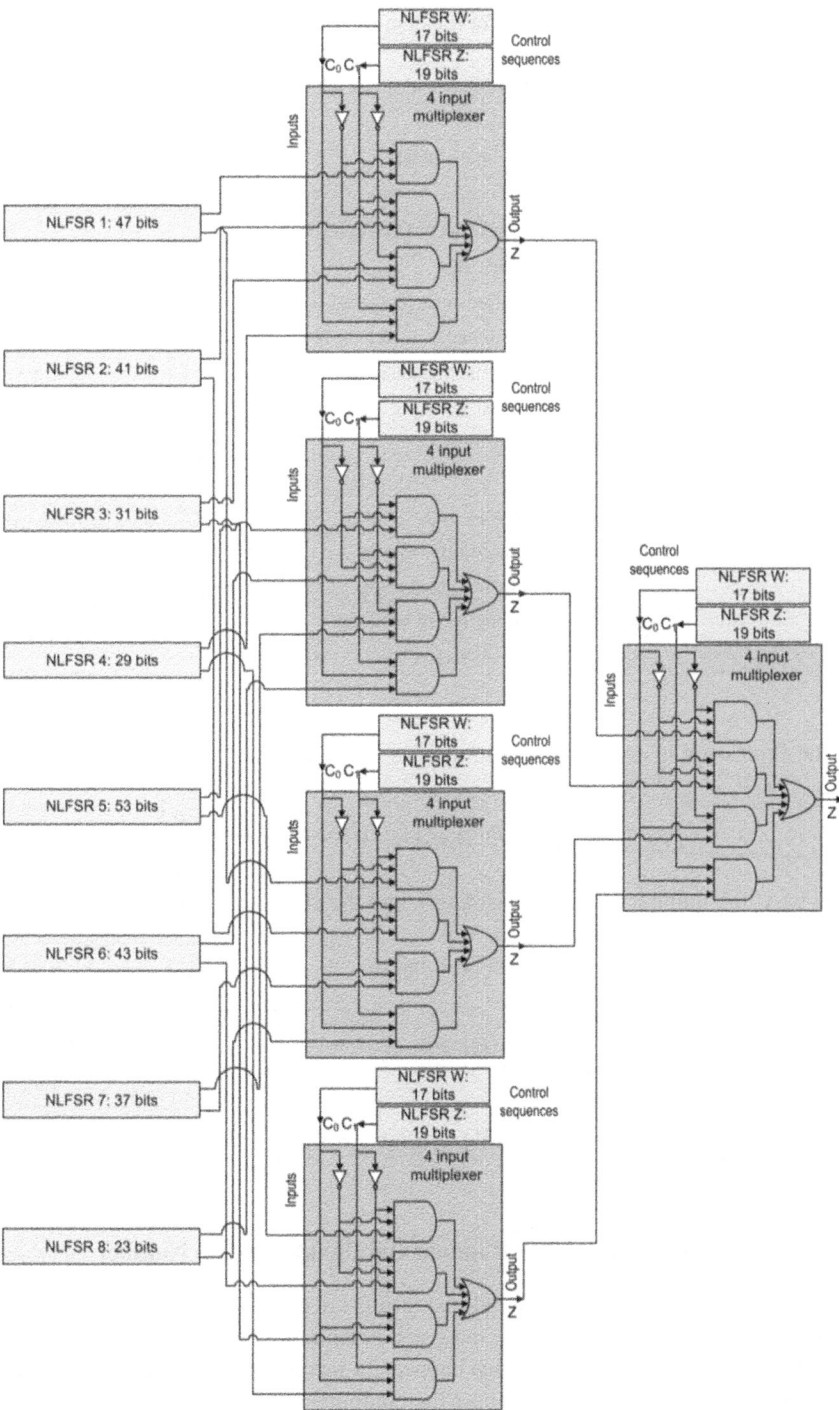

Fig. 5. Pseudorandom binary generator scheme with selected NLFSR

Table 5. Statistical Tests for Random and Pseudorandom Number

	Statistical Tests for Random and Pseudorandom Number
1	Frequency (Monobit)
2	Frequency Test within a Block
3	Runs Test
4	Test for the Longest Run of Ones in a Block
5	Binary Matrix Rank Test
6	Discrete Fourier Transform (Spectral) Test
7	Non-overlapping Template Matching Test
8	Overlapping Template Matching Test
9	Maurer's "Universal Statistical" Test
10	Linear Complexity Test
11	Serial Test:
12	Approximate Entropy Test
13	Cumulative Sums Test (Forward)
14	Cumulative Sums Test (Backward)
15	Random Excursions Test (8 subtests)
16	Random Excursions Variant Test (18 subtests)

The expression to determine the limits is as follows:

$$Upperlimit, Lowerlimit = (1 - \alpha) \pm 3 \cdot \sqrt{\frac{\alpha(1 - \alpha)}{k}}$$

In our case, the number of samples is: $k\,100$, and the chosen significance level is: $\alpha = 0.01$.

The upper and lower limits would be equal to:

$$Upperlimit = (1 - 0.01) + 3 \cdot \sqrt{\frac{0.01(1 - 0.01)}{100}} = 1.02$$

$$Lowerlimit = (1 - 0.01) - 3 \cdot \sqrt{\frac{0.01(1 - 0.01)}{100}} = 0.96$$

7.2 Statistical Tests (1 to 14) for Random and Pseudorandom Number

Once all the tests have been carried out, the results are placed in Table 6.

Using the data in the table above, the graph is built, then it is observed if the points are within the acceptance limits. If this happens, it means that the sequences produced by the binary generator successfully pass the randomness tests, in Fig. 6:

7.3 Random Excursions Test

Random Excursions Test, the subtests are executed and the results obtained are placed in Table 7:

Table 6. Statistical Tests for Random and Pseudorandom Number

	Statistical Tests for Random and Pseudorandom Number	Total	Pass	Prop	Upper	Lower
1	Frequency (Monobit)	100	100	1.00	1.02	0.96
2	Frequency Test within a Block	100	100	1.00	1.02	0.96
3	Runs Test	100	99	0.99	1.02	0.96
4	Test for the Longest Run of Ones in a Block	100	98	0.98	1.02	0.96
5	Binary Matrix Rank Test	100	98	0.98	1.02	0.96
6	Discrete Fourier Transform (Spectral) Test	100	100	1.00	1.02	0.96
7	Non-overlapping Template Matching Test	100	99	0.99	1.02	0.96
8	Overlapping Template Matching Test	100	100	1.00	1.02	0.96
9	Maurer's "Universal Statistical" Test	100	97	0.97	1.02	0.96
10	Linear Complexity Test	100	100	1.00	1.02	0.96
11	Serial Test:	100	100	1.00	1.02	0.96
12	Approximate Entropy Test	100	100	1.00	1.02	0.96
13	Cumulative Sums Test (Forward)	100	100	1.00	1.02	0.96
14	Cumulative Sums Test (Backward)	100	100	1.00	1.02	0.96

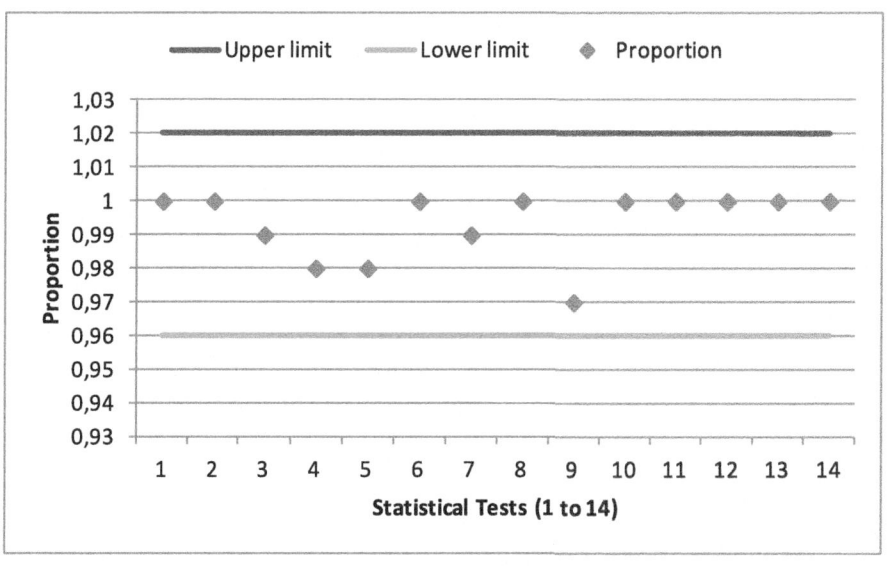

Fig. 6. Dot plot of Statistical Tests 1 to 14

Table 7. Random Excursions Subtests

	Statistical Tests for Random and Pseudorandom Number	Total	Pass	Prop	Upper	Lower
1	Random Excursions Subtest 1	100	99	0.99	1.02	0.96
2	Random Excursions Subtest 2	100	98	0.98	1.02	0.96
3	Random Excursions Subtest 3	100	100	1.00	1.02	0.96
4	Random Excursions Subtest 4	100	100	1.00	1.02	0.96
5	Random Excursions Subtest 5	100	99	0.99	1.02	0.96
6	Random Excursions Subtest 6	100	100	1.00	1.02	0.96
7	Random Excursions Subtest 7	100	100	1.00	1.02	0.96
8	Random Excursions Subtest 8	100	98	0.98	1.02	0.96

Using the data in the table above, the graph is built, then it is observed if the points are within the acceptance limits. If this happens, it means that the sequences produced by the binary generator successfully pass the randomness tests, in Fig. 7:

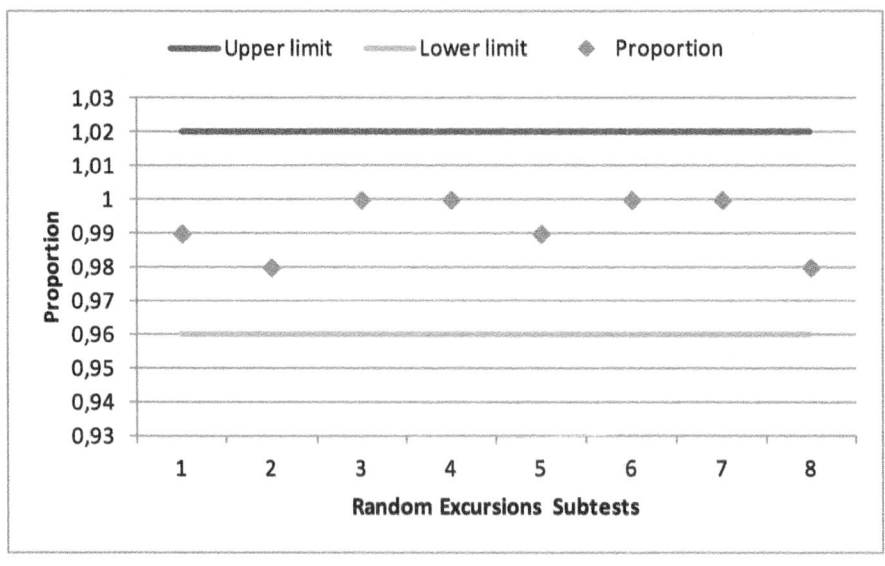

Fig. 7. Dot plot of Random Excursions Subtests

7.4 Random Excursions Variant Test

Random Excursion Variant Test, the subtests are executed and the results obtained are placed in Table 8:

Using the data in the table above, the graph is built, then it is observed if the points are within the acceptance limits. If this happens, it means that the sequences produced by the binary generator successfully pass the randomness tests, in Fig. 8:

According to the Tables 6, 7, 8 and Figs. 6, 7, 8, the hundred samples pass all the tests, which leads us to conclude that the generator produces effectively random sequences.

8   Concluding Remarks and Future Work

The pseudorandom binary generator presented in this paper combines in a non-linear way, using multiplexers, the sequences produced by eight NLFSR. The set of NLFSR produces eight binary sequences, which are mixed by four four-input multiplexers to obtain four pseudorandom binary chains, which are finally joined with another four-input multiplexer to obtain the final sequence.

The NLFSR that make up each generator have primitive connection polynomials, which ensure the maximum period in the resulting sequences, and non-linear boolean filtering functions.

The multiplexers are responsible for combining the sequences and ensuring the best cryptographic performance. Once the selection process was carried out, these devices were incorporated into the generator to then work with different key values and generate the respective binary sequences.

Table 8. Random Excursions Variant Subtests

	Statistical Tests for Random and Pseudorandom Number	Total	Pass	Prop	Upper	Lower
1	Random Excursions Variant Subtest 1	100	98	0.98	1.02	0.96
2	Random Excursions Variant Subtest 2	100	98	0.98	1.02	0.96
3	Random Excursions Variant Subtest 3	100	100	1.00	1.02	0.96
4	Random Excursions Variant Subtest 4	100	100	1.00	1.02	0.96
5	Random Excursions Variant Subtest 5	100	100	1.00	1.02	0.96
6	Random Excursions Variant Subtest 6	100	100	1.00	1.02	0.96
7	Random Excursions Variant Subtest 7	100	99	0.99	1.02	0.96
8	Random Excursions Variant Subtest 8	100	99	0.99	1.02	0.96
9	Random Excursions Variant Subtest 9	100	99	0.99	1.02	0.96
10	Random Excursions Variant Subtest 10	100	98	0.98	1.02	0.96
11	Random Excursions Variant Subtest 11	100	98	0.98	1.02	0.96
12	Random Excursions Variant Subtest 12	100	97	0.97	1.02	0.96
13	Random Excursions Variant Subtest 13	100	99	0.99	1.02	0.96
14	Random Excursions Variant Subtest 14	100	99	0.99	1.02	0.96
15	Random Excursions Variant Subtest 15	100	99	0.99	1.02	0.96
16	Random Excursions Variant Subtest 16	100	99	0.99	1.02	0.96
17	Random Excursions Variant Subtest 17	100	99	0.99	1.02	0.96
18	Random Excursions Variant Subtest 18	100	99	0.99	1.02	0.96

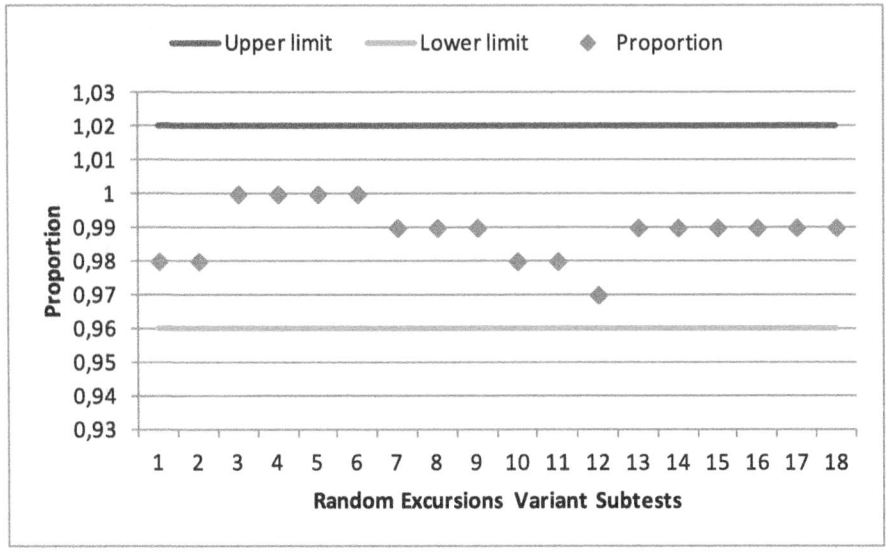

Fig. 8. Dot plot of Random Excursions Variant Subtests

Statistical tests of randomness and the subsequent interpretation of the results were carried out on them. The results obtained were satisfactory, so the presented generator is considered reliable for delivering pseudorandom binary sequences of excellent cryptographic quality.

In future developments, more NLFSR will be included and for non-linear filtering, four- and five-variable boolean functions will be used. In the fusion sector, all the resulting sequences will be combined with other joining devices. Finally, the final combination will be performed by a multiplexer whose number of inputs will be equal to the number of sequences it receives.

References

1. Jetzek, U.: Galois fields, linear feedback shift registers and their application. In: 17th International Symposium on Ambient Intelligence and Embedded System, Kiel University of Applied Sciences, Kiel, Germany (2018)
2. Lozano Cuevas, C.: Sucesiones de Recurrencia sobre Cuerpos Finitos y sus Aplicaciones, (Tesis). Universidad de Valladolid (2014)
3. Forner, E., Moreno, J.: Secuencias Pseudoaleatorias para Telecomunicaciones, Ediciones UPC (1996)
4. Fitzgerald, R.: A Characterization of Primitive Polynomials over Finite Fields. Southern Illinois University Carbondale (2003)
5. Brent, R., Zimmermann, P.: Algorithms for Finding Almost Irreducible and Almost Primitive Trinomials. Oxford University Computing Laboratory and LORIA/INRIA Lorraine (2003)
6. Gangopadhyay, S.: Boolean Functions in Cryptology. Department of Mathematics Indian Institute of Technology Roorkee. Tutorial Workshop on Many Facets of Cryptology (2011)
7. Musukwa, A.: Some cryptographic properties of Boolean functions, (Tesis). Universitá degli Studi di Trento (2019)

8. Tu, Z. and Deng, Y.: A Class of 1-Resilient Function with High Nonlinearity and Algebraic Immunity. Cryptology ePrint Archive, Report 2010/179. http://eprint.iacr.org/2010/179
9. Afflerbach, L.: Criteria for the assessment of random number generators. J. Comput. Appl. Math. **3**, 3–10 North-Holland (1990)
10. Rukhin, A., et al.: A Statistical Prueba Suite for Random and Pseudorandom Number Generators for Cryptographic Applications, National Institute of Standards and Technology (2000)

Author Index

GPSR Compliance

The European Union's (EU) General Product Safety Regulation (GPSR) is a set of rules that requires consumer products to be safe and our obligations to ensure this.

If you have any concerns about our products, you can contact us on ProductSafety@springernature.com

In case Publisher is established outside the EU, the EU authorized representative is:

Springer Nature Customer Service Center GmbH
Europaplatz 3
69115 Heidelberg, Germany

The manufacturer's authorised representative in the EU is Springer
Nature Customer Service Centre GmbH, Europaplatz 3, 69115 Heidelberg,
Germany. If you have any concerns regarding our products, please
contact ProductSafety@springernature.com

Printed and bound by CPI Group (UK) Ltd, Croydon, CR0 4YY

06/05/2026

02103601-0001